Fitness for work

The role of physical demands analysis and physical capacity assessment

Fitness for work

The role of physical demands analysis and physical capacity assessment

T. M. Fraser MB, ChB, MSc, PEng, FACPM
Former Professor and Chairman
Department of Systems Design Engineering
and Director
Centre for Occupational Health and Safety
University of Waterloo, Canada

Taylor & Francis

UK	Taylor & Francis Ltd, 4 John St, London WC1N 2ET
USA	Taylor & Francis Inc., 1900 Frost Road, Suite 101, Bristol, PA 19007

British Library Cataloguing in Publication Data

Fitness for work
A catalogue record for this book is available from the British Library

ISBN 0–7484–0018–4 (cased)
0–85066–858–1 (paper)

Library of Congress Cataloging-in-Publication Data is available

Cover design by John Leath

Contents

Preface vii

1 History and development 1
2 Traditional approach to medical examination 11
3 Job evaluation and analysis 23
4 Physical demands and analysis 39
5 Physical abilities analysis 67
6 Functional capacity assessment 81
7 Accommodations, restrictions and the handicapped 99
8 Job matching 107
9 Some legal considerations 119
Appendix A Physical demands job analysis 125
Appendix B Rating schemata for WCAM/WPAM analyses 159
Appendix C Physiologial and biomechanical techniques for work capacity measurement 165
Appendix D Physical demands and work capacity 177
Appendix E GULHEMP scale 183

References and bibliography 192

Index 209

Preface

Recently much attention has been directed towards the concepts of human rights, particularly in the Western world, and increasingly towards the needs of the underprivileged, the handicapped and the disabled in the industrial world. Some attention concerning the handicapped and the disabled has centred on the necessity for the reconsideration of medical selection procedures for entrants into the work-force, and particularly for those entrants whose potential capacities have been dismissed and whose limitations have been considered sufficiently paramount to deny them the dignity of work. Management, labour and government bodies alike increasingly realize that hiring of the less fortunate is a desirable objective, not only for humanitarian reasons, but also from sheer economic necessity.

Within this context, people are also realizing that previously conventional selection practices may no longer be tenable in the light of legislation about human rights and human privacy. Individuals have to be increasingly considered as in control of their own futures, and have to make and be held responsible for decisions which previously might have been taken almost arrogantly, and certainly unthinkingly, by an authoritative or paternal management.

Consideration of these and related matters led me to realize that not only is it desirable to re-evaluate medical and other selection procedures for the less fortunate, but also to reconsider the entire context of medical selection for all labour entrants.

Concomitant with realization of these considerations is recognition of the role and significance of 'physical demands analysis' and 'functional capacity assessment', the former pertaining to the demands of the work and the workplace, and the latter to the capacity of the worker to perform tasks within a working environment.

In this book I attempt to re-examine the whole concept of fitness for work, acknowledging the prerogatives of both management and labour, but emphasizing the necessity for consideration of humanitarian concerns and legislative requirements.

I should emphasize, however, that this is not a medical textbook. While certainly it should be of interest to members of the health professions in practice, in industry and in the educational institutions, I hope that it will be at least as useful to all those non-medical personnel who are concerned with the selection, placement and management of workers in the workplace, and to the business management and technical schools where these are trained. In addition, I believe it can be of value to all the policy makers and implementers from government, labour unions and other institutions where people seek to ensure a greater degree of equity in the practice and procedures of selection and placement.

The economics of personnel selection and placement has not been directly emphasized in this book. No chapter has been specifically oriented towards economics, but not the least reason for writing has been a recognition that, from an economic viewpoint, the placement of the right person in the right job must have a financial pay-off. As is suggested in the text, where the worker capacity equals the work demand there is significantly less lost time among workers than where worker capacity is either less than or greater than the work demand. Similarly, using handicapped persons in meaningful and productive tasks within their capacity is economically more desirable than providing them with some often inadequate support from the public purse.

This book is a collation and a crystallization of ideas and practices from a number of sources, some my own, but many from others who have been thinking and developing their work along the same lines. I hope that in bringing these concepts together I may have introduced a little coherence and systematization into what otherwise has been a somewhat diffuse topic, with literature ranging over a wide field and embracing the social and health sciences, the physical sciences and the arts and practices of human resource management. If I have failed to do so, or if I have made confusion doubly confused, then the blame lies with me, not with my sources.

T. M. Fraser

Chapter 1
History and development

Background

Recently increasing awareness of social issues has led in various jurisdictions in the USA, Canada, Australia, New Zealand, Great Britain, Scandinavia and some European countries to the enactment of legislation for the employment of the handicapped. This in turn has led to the recognition of the inadequacy of the procedures used for the medical evaluation of potential employees, whether handicapped or not. In some respects the procedures are misleading or even useless, as they are often based on a form of physical or medical examination which pays scant attention to the needs of the workplace. The laying on of hands has been a medical ritual at least since the days of Hippocrates. Physicians have dutifully applied stethoscope to chest and hands to belly, regardless of whether the subject was sick, well, wanting insurance, seeking employment or returning to work after illness or injury. Usually the questions, examinations and objects have been the same, namely, to compare the status of the person being examined against some vaguely defined concept of clinical perfection.

The potential employer, equally, has paid little attention to the actual requirements of the job or tasks and, with due regard to the 'bottom line', has often been more concerned with ensuring that his future compensation costs would not increase than with determining who would be most appropriate for the task at hand. In consequence, situations have sometimes arisen where employees have been hired whose physical and mental capacities exceed or fall short of the demands of the job, or perhaps worse, potential employees who could have performed specific tasks have been rejected because of failure to meet some ill-defined standard.

Out of these and related considerations has arisen a new approach to the evaluation of potential employees based on the concept of job matching, that is, comparing the functional capacities of the individual against the physical demands of the job. Although the concept is beginning to be appreciated, and the approach applied, there is little in the way of definitive knowledge among those most in need of it. There is, however, an increasing demand in industry, in the teaching institutions and within the health professions for information as to the nature of the concept and the methodology of its application. The sparse literature available is widely distributed in journals about industrial psychology, occupational health, rehabilitation, ergonomics, kinesiology and personnel and human resources research, along with occasional articles in trade and business magazines, and even more esoterically in the limited publications of sundry institutes, government departments and the like.

It is the intent of this book to bring this material together in readable form and to provide an understanding of how to use the methodology in actual practice and what might be expected from its use.

Origins of occupational medical examinations

The earliest known practice of physical examination of the person would appear to have been undertaken by the Ancient Egyptians, as illustrated in their hieroglyphic records. By the time of Hippocrates, who even advocated consideration of potential industrial exposure, it was well established. Knowledge of what happened about physical examinations between the time of the Egyptians and Hippocrates, however, is scant. It is clear that in the 400 years between Hippocrates in Greece and Galen in Rome much of what had been developed was again lost, and that physicians, perhaps like some of their telephone-oriented counterparts today, tended to evaluate the health of their patients in the absence of any direct contact. Indeed, the whole art and practice of medicine largely disappeared into a haze of herbal mysticism on the one hand and barbaric surgery on the other until well into the 15th century when, spurred by new ideas infiltrating from the highly developed systems of Arabic medicine, the great medical schools of Western Europe began to be formed and dispense their new ideas of scientifically based medicine. It took another 50 years, however, for that bizarre genius Paracelsus to appear on the scene and once again orient medical thought to direct observation of nature and the patient, and to insist that the key to evaluation of sickness and health lay in examining the person, and not in mere theoretical speculation. Although despised and ultimately condemned by his peers, his influence has been profound and many of his concepts remain today.

By the end of the 18th century a new influence was making itself felt, this time on the other side of the balance, as it were. The Industrial Revolution, which began in England in the late 1700s, was pulling thousands of previously agricultural workers or minor craftsmen into new and burgeoning industrial workplaces where little concern was felt or expressed for their health, safety or social conditions. The machine was sacrosanct; the worker was expendable. Little or nothing was done to evaluate their fitness for work other than establishing their willingness to do the task demanded and the apparent absence of any limiting disability.

Medical significance of 19th century social reform

The early 19th century in Britain saw the introduction of the British Factory Acts under the direction of the great humanitarian statesmen and industrialists Sir Robert Peel and Robert Owen, along with the physicians Sir John Simon and Thomas Southwood Smith. The Factory Acts, of which there was a series beginning in 1833 and continuing through until 1867, as well as the writings of such authors as Charles Dickens and Charles Kingsley, drew attention to industrial working conditions and mandated at least the beginnings of control programmes. Indeed, by the Act of 1844 the requirement for physcial examinations of workers was initiated in selected industries.

However, despite the undoubted reforms brought about by the Factory Acts, physical examination of the potential worker was still rudimentary, when it occurred at all. Other than in the specific legislated industries, any physical examination tended to be a cursory inspection by the employer or his representative and was largely limited to determining whether a potential worker, impaired or not, appeared to have sufficient capacity to do the job at hand. Furthermore, if a worker became injured or sick because of the nature of his work or working conditions he had no recourse except through the civil courts, a route that was rarely available to an impecunious and out-of-work employee.

Role of social insurance and workers' compensation

At the end of the 19th century following the development of social insurance schemes by Bismarck in Germany, a new piece of legislation was introduced in Britain which ultimately influenced the entire industrial world, and, perhaps unwittingly, determined the future needs for physical examination of the person. This was the Workmen's Compensation Act of 1897 which provided for automatic compensation to an injured worker except in the case of wilful misconduct. The original Act was limited in scope to a few industries and situations, but was expanded in 1907 and ultimately became a world model. New Jersey and Wisconsin were the first American states to follow, and Ontario became the first in Canada in 1920. The other English speaking countries, along with Japan, the Scandinavian countries, and ultimately the whole industrial world, developed similar schemes of government funded workers' compensation or private no-risk insurance until the concept became virtually universal in industrialized countries. Recently the scope of the concept has expanded to include certain occupational sickness as well as injury.

While workers' compensation acts and equivalent insurance schemes were a great step forward in social legislation, their presence introduced a new and unexpected problem for the worker, and particularly the impaired or handicapped worker. The funds to provide the compensation are derived from a levy on employers based on the relative expected risk of work in a given industry, and the injury record of the specific employer in that industry. Thus, employers with a high injury record are required to pay at a higher rate than employers with a low injury record. Recognition of this differential led some unscrupulous employers in the early part of the 20th century to attempt to exclude from their employment any employee who was impaired, incapacitated in any way, or who had a record of sickness or chronic illness, regardless of whether or not it had or would have any effect on his/her capacity to work. Indeed, the pre-employment physical (medical) examination was initiated in reality for this purpose. Because of the demand on the part of these employers to minimize the financial burden of workers' compensation, the emphasis of the examination in these situations became one of definition of problems, diagnosis of sickness and incapacity, exclusion of contagious disease, and rejection of all but the fit. As a result, many potential workers who could have made valuable contributions to the work-force were excluded, to become a burden to themselves, their families and the social services.

This approach continued at least for the first few decades of the 20th century, although Mock (1920) showed an early recognition of greater responsibilities when he wrote: 'Every applicant for work should be thoroughly examined by the medical staff in order to prevent the introduction of contagious diseases into the plant and to provide for the *proper selection of work for every man according to his physical and mental qualifications*' (present author's italics).

Development of systematic health-oriented medical examinations

The beginning and development of World War II from 1939 and into the 1940s, with its concomitant need for fit and healthy members of the armed services, saw a focused entry or re-entry into the civilian work-force of females, as well as workers who were older and less fit. Even then, however, little provision was made for their evaluation. In the case of the military services, at least initially, very little more was required than to be in an eligible age group, without chronic disease or obvious impairment, before some placement could be found, while for civilian work in most cases all that was required was a demonstrated willingness to work.

The American Medical Association was one of the first groups in organized medicine to recognize the need for a systematic approach to physical examinations in industry. In 1944 a special report on such examinations was submitted to the Association's governing body. In this report it was noted that health examinations should be considered as a means towards achieving the promotion and maintenance of the physical and mental health of workers in industry, and that unjust exclusion from work, or exclusion on doubtful grounds through the improper application of findings from such a health examination, was against public welfare and contrary to sound industrial health principles.

These considerations and others resulted in the formation of the (US) War Manpower Commission, also in 1944, whose role was to plan programmes to use workers effectively through the appropriate allocation of people and skills. During and preceding this period the U.S. Civil Service Commission had made a classification of all disabling conditions and paired these states with compatible positions available in federal employment. The initial classification of these positions ultimately became the *Dictionary of Occupational Titles* (US Department of Labor, 1977).

The method initially used by the Commission in the implementation of its mandate was to compare each of 493 positions in a Richmond, California, shipyard against a checklist they had devised. The checklist was intended to encompass the type of generic activities that would be encountered in the course of various tasks, and indeed represented the first attempt at defining work activities on the job. The descriptors included definition of the requirements for standing, crouching, sitting, walking, climbing or throwing, as well as sensory demands such as feeling, hearing and colour vision, and the need for speech. No consideration was given at that time however to visual disabilities, nor was any accommodation made for workers in wheelchairs. The checklist was further developed later to include consideration of adverse environments.

Further extension of the concept by the US Department of Labor, as mentioned earlier, led to the analysis of a very wide variety of jobs in terms of strength requirements, ranging from sedentary to very heavy, and including a significant need for climbing or crawling, a significant need for reaching, handling, fingering and feeling, as well as for talking, hearing and seeing.

Definitive work, however, is associated with the name of Dr B. Hanman (1945, 1946, 1948, 1958) at the Kaiser shipbuilding facility in San Francisco and the US Naval Air Station in Alameda, California. In a series of papers beginning over 40 years ago, the importance of which was then underestimated, Hanman pointed out many of the limitations that can arise from the use of the traditional approach to medical (physical) examinations as a means of determining the fitness of a worker to work, and initiated many of the concepts of physical demands analysis and functional capacity assessment which are considered in this text. Hanman's work is discussed in detail in a later chapter.

It is useful to recognize that Hanman classified the traditional approach in two categories, namely the 'rating method' and the 'disability method'. In the rating method, which is the most common, adjectives and adverbs are used to describe the extent of a person's ability to undertake activity, such as good/better/best, little/moderate/great, and occasionally/frequently/constantly, or as another example, 'no heavy lifting'. He notes, however, that while these phrases may have specific meaning to the originator, they may have a different meaning to the employer, or even the patient. Does the phrase 'no heavy lifting' mean moderate or little lifting, and if so what is meant by moderate or little. Does it refer to the weight, the frequency, the duration of holding and carrying, or what? Even vaguer examples occur such as 'fit for light work', or 'light duties only'. Indeed, Hanman (1958) points out that according to how one interprets the scale definitions developed by the US Civil Service Commission, a worker classified as 'fit for little lifting' could be required to lift 40 lb at a time for 3 h per day, while a person classified as 'fit for moderate lifting' would not have to lift more than 15 lb for 3 h, and a person classified for 'great lifting' would not have to lift over 45 lb per day.

He describes the disability rating method as occurring when jobs and tasks are defined as being suitable for persons with specific disabilities, such as the absence of a limb, or chronic asthma, in the belief that since disabled persons in any specific disability group are alike in their disabilities they are also alike in their abilities — a belief which of course is erroneous since disabled persons vary in their capacities just as much as those who are not disabled. A second error occurs when a disability group is considered in terms of 'average' performance, which ignores the actual capabilities of the individual person. The disability rating method thus accentuates the negative aspects of work capacity instead of determining the positive aspects of the worker's ability.

Recognizing the limitations of each of these methods Hanman (1958) continues:

> Through subjectivity and misunderstanding, the lives of certain persons will be unduly restricted because they avoid activities and others restrict them from activities that they very well could perform with safety. To thousands of people this kind of misunderstanding means the difference between getting a suitable job or not and between living a more enjoyable life of retirement or not. But, still worse, certain other persons

will themselves undertake activities, and will be allowed by others to undertake activities, that they cannot perform with physical safety.

Hanman goes on to define what he calls the specific method for medical evaluation which is the foundation of functional capacity assessment, and will be examined later.

Echoing Hanman's comments, and in the light of her own experience, Slavenski (1986) outlines some of the reasons for failure of the traditional selection process as follows:

— it is not based on an analysis of job requirements;
— rather than being structured and systematic, it is informal and inconsistent, making it difficult to compare and evaluate candidates;
— it may involve irrelevant, and sometimes illegal, questions;
— it allows candidates little opportunity to demonstrate actual skills;
— it may be based on poor observation and documentation and usually relies on the interviewer's ability to recall complex information about a number of candidates.

Furthermore, as Abt Associates (1984) point out in their study, candidates and their families can be affected economically and emotionally by adverse findings. Workers barred from the workplace by reason of alleged unfitness cannot contribute to the productivity of society, while society may spend substantial sums on rehabilitation, insurance benefits and compensation, based on what may be an inadequate assessment of their capacity to work. Thus, in considering fitness for work, one has not merely to recognize whether an applicant is 'normal' or not, but whether in fact an abnormality constitutes an impairment or handicap.

It becomes necessary, then, to define the terms impaired, disabled and handicapped. The World Health Organization in its *International Classification of Impairments, Disabilities and Handicaps* (WHO, 1980) defined them as follows:

Impairment: In the context of health experience, an impairment is any loss or abnormality of psychological, physiological, or anatomical structure or function.

Disability: In the context of health experience, a disability is any restriction or lack (resulting from an impairment) of ability to perform an activity in the manner or within the range considered normal for a human being.

Handicap: In the context of health experience, a handicap is a disadvantage for a given individual, resulting from an impairment or a disability, that limits or prevents the fulfilment of a role that is normal (depending on age, sex, and social and cultural factors) for that individual.

While these definitions may be valuable from a conceptual viewpoint, it is desirable for practical purposes to have a more operationally oriented definition. According to the US Department of Labor Regulations, a handicapped individual is defined as one who:

(a) has a physical or mental impairment which substantially limits one or more major life activities;
(b) has a record of such impairment;
(c) is regarded as having such impairment.

In this connection, physical or mental impairment is considered to mean (a) any physiological disorder or condition, cosmetic disfigurement, or anatomical loss affecting one or more of the following body systems: neurological, musculoskeletal, special sense organs, respiratory, including speech organs, cardiovascular, reproductive, digestive, genito-urinary, haemic and lymphatic, skin and endocrine; and (b) any mental or psychological disorder, such as mental retardation, organic brain syndrome, emotional or mental illness, and specific learning disabilities.

The term 'physical or mental impairment' is considered to include, but is not limited to, such diseases and conditions as orthopaedic, visual, speech and hearing impairments, cerebral palsy, epilepsy, muscular dystrophy, multiple sclerosis, cancer, heart disease, diabetes, mental retardation, emotional illness and drug addiction and alcoholism.

The term 'substantial limits' is considered to refer to the degree that the impairment affects an individual becoming a beneficiary of a programme or activity receiving federal assistance or affects an individual's employability.

Bearing these considerations in mind it should be recognized that in most industrialized societies management is no longer permitted to discriminate in recruitment, hiring, compensation, job assignment and classification, and fringe benefits, against any applicant for employment on the grounds of impairment, or because the applicant's condition might lead to a shortened working life, premature retirement on grounds of disability, early death, or a drain on a life insurance programme, or excessive need for medical care in the workplace. As will be discussed later, the employer in fact is required to make 'reasonable accommodation' on behalf of such applicants unless it can be demonstrated that the accommodation would impose an undue hardship on the employer. Indeed, again in the more advanced jurisdictions, not only do all testing procedures need to be job related, but pre-placement examinations must also be oriented to the job or jobs available.

Job matching

From the foregoing then it must become evident that some form of specific job matching is desirable for the needs of both society and the worker. As a member of the Ontario Human Rights Commission has remarked (Ramanujam, 1988):

> To this day, some employers seek to establish an extensive profile on individuals, especially in the areas of health; even for the most innocuous jobs individuals are often asked to provide not only a medical biography but a history of diseases in their family, and are required to undergo extensive medical exam. For those who have diseases or conditions that are still largely misunderstood by the public — for example, diabetes, epilepsy, cerebral palsy, and many psychological disorders — this process of disclosure is fraught with risk, since the process of disclosure all too often ends with the mere identification of the disease or condition and does not proceed to determine the severity of it, whether it has anything to do with the individual's ability to do the job, and whether there is any way the individual can perform the essential duties of the job despite the health problem.

Human rights legislation has evolved in response to the increasing recognition of the potential for abuse in the traditional practice of obtaining information from employees and applicants. Under various forms of Human Rights Codes, as practised in different states, provinces and countries, employers are prohibited from obtaining information in the application and recruitment stage of personnel hiring that may classify a worker by a prohibited ground of discrimination. An employer may enquire into a person's physical and/or mental health status only when those enquiries are necessary to determine whether the job can be performed safely and efficiently. This is equally true of the application and interview stages of the recruitment process, and of the assessment for return to work after injury or illness. These matters will be considered further in a later chapter.

Thus, in the light of Code requirements, and specifically the limitations imposed on the gathering of information, a need to establish a more accurate match between the individual and the particular demands of the job has arisen. This in turn means that during the recruitment process a management professional will be required to make a decision based on the evaluation of medical information and job analysis information. The provision of medical information is predicated on the assessment of the functional capacity of the applicant by a health professional or professionals, while provision of job information is dependent on the evaluation of knowledge derived from an incumbent and/or a formal and systematic job classification system. Thus, the concept of job matching has two major components, namely a physical demands analysis (PDA), and a functional capacity assessment (FCA).

Physical Demands Analysis (PDA)

A physical demands analysis differs from the usual job description (see later) in that it details the physical requirements of job-related tasks. It identifies such physical requirements as lifting, walking, bending, writing, reading, typing, and so on. A simple format can be used as a routine personnel practice to classify and describe positions. More elaborate assessments of PDA can be done to identify accommodations, aids, and so on, to enhance the productivity of a disabled individual. The concept includes that of physical abilities anlaysis. These concepts and techniques will be elaborated in Chapters 4 and 5.

A PDA should not be confused with a standard job description. A standard job description is commonly intended for determining the job/wage classification of a worker and for defining the tasks usually undertaken in a particular job. The PDA requires a much more definitive and precise approach, outlining the specific requirements for such modalities as arm strength, back mobility, endurance, vision, dexterity, and so on.

Functional Capacity Assessment (FCA)

Functional capacity assessment, sometimes known as functional ability evaluation, is used to examine the capacity of a person to perform the essential duties of a position. It is of necessity conducted in the light of a previously completed PDA. FCA is not one technique, but a range of assessment procedures used to evaluate fitness for work. It provides a basis for matching an individual to a specific job or general type of work, and may or may not include suggestions for aids and accommodations to improve the match. FCA may be done routinely for all employees, or in particular cases, such as in assessing the readiness of a previously ill or injured employee to return to work.

FCA differs from a traditional medical examination primarily in its orientation and to a lesser extent in its content. In a traditional examination the question of whether an applicant is fit for specific work, or is capable of performing the essential duties of a job, is frequently not fully addressed, In an FCA consideration of the nature of the work and other factors such as the person's skills, motivation, and so on, is paramount. A person may be fully fit to work at a specific task in one setting, but incapable of work at a different task in another setting. Unless the tasks and settings are known, the fitness cannot be properly assessed. Both PDA and FCA will be examined in detail in later chapters.

The approach to job-matching thus becomes a positive one. It emphasizes what a person is capable of doing. It may specify aids and accommodations, but it is only complete when the person concerned has been demonstrated to be successful in the job or incapable of performing the essential functions of the job even with counselling, aids and, if necessary, a job trial.

Chapter 2
Traditional approach to medical examination

Medical examinations, or physical examinations, as they are known in the USA and Canada, can be categorized into two basic types, namely, examination of the patient with a complaint, and examination of the apparently well person. Each is purpose-oriented, but the purposes are very different.

In the first case, examination of the patient with a complaint, the examination is in the form of a diagnostic clinical review with a follow-up if necessary, and is problem-oriented. The physician or surgeon who normally conducts the examination is concerned with the definition of the problem(s) propounded by the patient and identification of the probable cause(s) with a view to removing or alleviating any adverse condition found. In a subsequent examination, the physician/surgeon may also be concerned with evaluating the progress of any such condition and the effects of any therapeutic intervention.

The examination will normally include a brief general medical history, minimal when the patient is known to the examiner, and an extensive history pertinent to the problem(s) at hand. Similarly, the manual examination will tend to be oriented to the specific complaint, and will only be extended when circumstances indicate a need. Laboratory and other special facility examinations will also be problem-oriented. The physician, indeed, is not concerned with assessing the general fitness of the patient, except in so far as it impinges upon the problem, nor is he/she concerned at that time with the specific fitness of the patient to work or to do a particular type of task. These considerations may be assessed after the problem(s) at hand have been dealt with, but then they tend to be assessed in the light of the findings of the problem-oriented examination, which of course is concerned with definition of incapacity and not functional capacity. Furthermore, the examiner has sometimes little appreciation of the demands of the work other than what can be gleaned from the patient.

Examination of the apparently well person is not problem-oriented in the same sense. The examination is purposeful in that the examiner has some objective in mind; normally to determine as far as possible the suitability of that person for a particular purpose, the demands of which have been provided with variable clarity by some outside source. These demands may define such characteristics as suitability for life insurance, for acceptance by various branches of the armed services, for general employment by a company, for employment in a particular type of task, and so on. The applicant does not present with a health problem, although such a problem may be revealed in the course of the examination. The applicant assumes that he/she is healthy, or may indeed attempt to conceal an incapacity or impairment. The function of the examiner is to define the state of applicant's fitness by some standard and to attempt to predict the applicant's future health by the probability of incapacity to come. In the traditional form of fitness examination, just as in the diagnostic examination, although the purpose may be different, the orientation again is towards defining or predicting present or future incapacity, and not in assessing functional capacity, nor in suggesting any reasonable changes that might be made in the workplace to accommodate an acceptable incapacity.

Nature of the fitness-for-work examination

Before discussing the concepts and practice of physical demands analysis and functional capacity evaluation in later chapters, it will now be useful to outline the development and practices of the traditional medical examination for fitness for work.

The fitness-for-work examination, as the name implies is an example of the previously mentioned examination of the apparently well person. It goes by various other names depending on its purpose, such as the pre-employment examination, the pre-placement examination, and the periodic examination. As a concept, the pre-employment examination, which is given to a worker prior to his being offered a job in the company concerned, has largely disappeared, partly because in many jurisdictions such an examination is considered to be an infringement of human rights, and partly because it is wasteful in effort and money to examine a person who may be rejected at an early stage in the employment process for other reasons. The pre-placement examination is still a commonly used term and refers to a medical examination that is conducted after a potential employee has accepted an offer of employment conditional on his medical suitability, while the periodic examination is a regular examination of workers for the purpose of detecting health conditions potentially related to work, which may or may not be treatable in the individual worker, but might be treatable in the public health sense by recognition and control of hazards in others (Halperin *et al.*, 1984).

Objectives of the fitness-for-work examination

Numerous authors have discussed the purpose, needs, content, usefulness, validity and limitations of pre-placement examinations, beginning with Sawyer in 1942 who based his concepts of fitness for civilian work on the concepts of the military and insurance medical examinations of his time, and including others such as Alexander *et al.* (1975), Brubaker (1972), Campione (1972), Catalina (1983), Cowell (1986), Goldman (1986), Hanks (1962), Harte (1974), Hogan and Bernacki (1981), Kelly (1965), Lincoln (1968), Luongo (1962), Michaels (1973), Present (1974), Shevick (1961), Schender and McDonough (1971), Schussler *et al.* (1975) and Todd (1965). Examinations for military duties dominated the scene during World War II, and indeed little in the way of analysis of physical examinations is reported in the literature until the 1960s. Meantime, however, routine relatively unstructured pre-employment, and even pre-placement examinations, were becoming a necessary part of the employment process, at least for the larger companies, although more and more persons were beginning to doubt their usefulness.

It was not until the 1970s that any definitive attempt was made to outline the objectives of a fitness-for-work examination. In 1973, the American Medical Association (AMA, 1973) defined some guiding principles for the conduct of medical examinations in industry. They stated that the objectives were to:

(a) measure the medical fitness of individuals to perform their duties without hazard to themselves or others;
(b) assist individuals in the maintenance or improvement of their health;
(c) detect the effects of harmful working conditions and advise corrective measures, and;
(d) establish a record of the condition of the individual at the time of each examination.

It is interesting that even then there was included in the guidelines a specific statement that emphasis should be on the placement of individuals according to their abilities and not simply selection of the physically perfect with rejection of all others — a tacit recognition of the need, not always effected in practice, to define human capacities with the intent of matching the person to the job, rather than to determine human liabilities for the purpose of eliminating the imperfect.

The AMA guidelines crystallized the best practice that had developed over the previous few decades, and indeed provided guidelines which are still mainly in use by medical practitioners today.

The primary goal of the routine pre-placement examination, then, is to reveal any medical condition that might put the worker in a situation of increased risk to himself or others. Additionally, the examination may provide a baseline for future testing to assess the impact of subsequent exposures or development of pathological conditions (Goldman, 1986), and to ensure that the employer meets any statutory liability (Harte, 1974). Weed (1969) also pointed out that the pre-placement examination is the point in an employee's career where a data base can be established and where a list of health problems can be defined for each employee.

Clinical methodology

While most practitioners have developed routines of general physical examination, gleaned from early training, experience and many different sources, certain specific requirements have been suggested by various authorities for the conduct of occupational health examinations. The original *Guidelines* of the American Medical Association (AMA, 1973) emphasize flexibility and note that what may be adequate in one case may be insufficient in another, and that it may indeed be economically unsound for an essentially non-hazardous industry to routinely include for instance, extensive laboratory and X-ray testing.

The *Guidelines*, however, do suggest that as a minimum an examination should include general and relevant medical history, age, height and weight, along with consideration of general appearance, skin, eyes, ears, nose, teeth and mouth. Routine examination of pulse, blood pressure and temperature is suggested, along with examination of chest, lymph nodes, abdomen (including hernia), anus, genitalia, spine and extremities. Urinalysis, vision and hearing testing are also part of the test procedure. They also observe, without comment or interpretation, that 'significant nervous or mental manifestations should be noted'.

Whether or not such an examination has any meaningful validity remains to be considered in a later section, but the principles of that type of examination are entrenched even more deeply, and with even greater comprehensiveness, in an article by Catalina (1983) in the prestigious *Encyclopedia of Occupational Health and Safety* in which he details the requirements of an occupational health examination as follows:

Physical characteristics
Weight, height, chest measurements, dynometric performance of the most frequently used muscle groups (lumbar, scapulodorsal, manual), sturdiness of applicant, stature.
Sensory characteristics
Visual acuity, near and distant, field of vision, oculomotor function, colour vision, visual defects, hearing (watch, tuning fork, audiometry), sense of touch.
Function testing
Locomotor apparatus, e.g. abnormalities of spine, lower limbs; capacity for walking, bending, stretching, rotation; abdominal wall (hernia); skin (occupational skin disease); upper respiratory tract.
Major systems
Urinalysis for sugar, albumen, blood; cardiovascular system — blood pressure, heart rate, auscultation, examination of peripheral pulse, venous system, particularly in lower limbs (varicose veins); respiratory system by auscultation and radiology; digestive system by history, palpation (alcoholism, diabetes, peptic ulcer, absorption of toxic substances); neural disorders of limbs; genital apparatus, pregnancy, hormonal disorders, splenomegaly.

While the AMA *Guidelines* present what were regarded as minimal requirements for their time, it would seem that Catalina's approach to comprehensiveness is over cautious, perhaps to ensure that the examiner cannot be accused later of failing to meet professional obligations. Most traditional examinations fall somewhere in between the two outlines. A more useful approach will be discussed when examining the concept of functional capacity analysis.

The questionnaire as an examination tool

There is little doubt that the occupational health history is the cornerstone of an occupational health examination (Goldman, 1986). The question of whether a properly constructed questionnaire can either replace, or perhaps form the most significant part of, an examination has been examined by Harte (1974), and also by Schussler *et al.* (1975).

Harte (1974) conducted a study within a hospital environment to establish the most effective procedures for assessing the health status of employees. As an aside, he noted that medical staff without exception wanted others to be examined, but not themselves, and in particular they all expected student nurses to be examined, an observation which Harte wryly notes '. . . may reflect the anxieties of employment in a stressful occupation or possibly just that doctors like examining nurses'. He also

Table 2.1. Section of pre-placement examination responsible for restrictions (Schussler, 1975).

	1st study		2nd study	
	No. of findings	%	No. of findings	%
History	47	54	135	36
Basic measures	28	32·2	180	48
Medical examination	9	10·3	51	13
Nurse observation	3	3·5	12	3

noted — an observation that is widely held — that employees feel that the medical examination is not really conducted for their benefit, that it is unnecessary and that it may be a threat to their employment.

Accordingly, he compared the information obtained from a specially developed health questionnaire administered by nurses and accompanied by an interview, with that obtained by a traditional medical examination. He found a considerable similarity between the two, and went on to say:

> I believe that if the medical examination is looked upon as a routine procedure for staff selection it will be barren in results, frustrating to the doctor and the employee and of little value to the employer, but if it is a dynamic problem-oriented record it can provide a data base for health surveillance and a record for subsequent evaluation. I am certain that selective full medical examinations are necessary for a limited number of employees who either work in 'at risk' jobs or are vulnerable as persons, but in most instances the health evaluation is best done by a questionnaire and health interview by the occupational health nurse who has the key role in the effective and efficient occupational health services.

Schussler *et al.* (1975) also conducted a comparative study which involved the evaluation of findings on pre-placement examinations of nearly 3000 persons in one instance and of over 2000 in another. The examination included basic measurements and visual observations conducted by a nurse such as height, weight, blood pressure and pulse rate, as well as urinalysis, blood haemoglobin, visual acuity, chest X-ray and physical examination by a physician. It was necessary to apply restrictions to some 14·5% of these applicants. Of that group, the portion of the examination that could be credited for providing information on which to base the restrictions is shown in Table 2.1.

Although it is certainly clear that in the first study restrictions could be applied on the basis of information from history alone, the more striking feature is that in both instances relatively little additional information was gained by a full medical examination.

X-ray examination

Because there is little of value to be gained, and to reduce to a minimum exposure to ionizing radiation, the use of routine chest X-rays has largely been discontinued except for employees in special situations such as silica workers, isocyanate workers and others where there is a real possibility of hazardous exposure. Routine chest X-rays should not in fact be used except where there is justifiable cause.

The question of routine back X-rays, however, remains somewhat controversial, although again consensus appears to favour their abandonment. During the 1950s and 1960s various clinicians recommended lumbar X-rays as part of a low-back injury screening programme (Becker, 1955; Crookshank and Warshaw, 1961; Kelly, 1965; McGill, 1968). Kelly (1965) in particular investigated over 1000 applicants for employment who had received lumbosacral X-rays, and found that 3·4% showed defects sufficient to cause their rejection, while 13·8% showed defects which permitted employment only with limitations. Crookshank and Warshaw (1961) compared two sets of data. In one group of 1927 candidates following lumbosacral X-ray, 1503 were placed in jobs suitable for their apparent status. Of these only seven persons sustained non-traumatic back strain during a subsequent five-year work period, with none requiring work loss. In contrast, among 3395 workers in similar jobs who did not receive lumbosacral X-ray, 254 cases of non-traumatic back disability occurred during the same period.

On the other hand, Chaffin (1975), quoting various epidemiological studies to substantiate his own extensive experience, considers that the use of routine lumbosacral X-rays has not been validated as the sole criterion for employability. He points out that the major interpretation problem appears to arise because the radiological examinations are most often simply a part of a total preventive programme. Hence, while the statistics pertaining to incidence of back strain may show an improvement in the number and severity of cases after the initiation of such a programme, the results could just as readily arise from the general use of more rigorous screening processes. He goes on to state:

> It is, therefore, necessary to ensure that when a specific positive radiological attribute is used to limit a person's employment opportunities, a good biomechanical basis is documented to relate the skeletal defect to the stresses to be expected if the person performed the job.

Medical categories and disability scales

In the traditional form of medical examination for fitness for work, once a physical examination has been completed it is necessary to classify the health status of the examinee in such a manner as to permit the employer to determine his employability. The earliest users of classification systems were the military. Their approach will be discussed in some detail later, but essentially their classification systems involved some form of coding indicating that a candidate was fit or unfit for various kinds of military service.

Most civilian classification systems in use today are based with various modifications on the original work of Sawyer (1942) who defined a scale as shown in Table 2.2.

Luongo (1972) studied some 7500 pre-placement examinations from 1948 to 1962

Table 2.2. Fitness classification schema originally developed by Sawyer (1942).

Class	Description
A	Physically fit for any work
B	Defect negligible or correctable; otherwise fit for any work
C	Defect limits fitness for certain work and/or requires medical control
D	Defect requires medical control or attention and disqualifies for employment

Table 2.3. Fitness classification system suggested by the American Medical Association (1973).

Class	Description
A	Physically fit for any work
B	Defect that limits fitness for work and leaves applicant eligible for certain jobs (the defect may or may not be correctable but may require medical supervision)
C	Defect that requires medical attention, is presently handicapping, and disqualifies for any type of employment

Table 2.4. Health criteria as suggested by Lincoln (1968)

Class	Description
1	Exceptionally good health
2	Good health but has minor abnormalities which should not significantly affect future health prospects
3	Average health — has minor abnormalities or borderline findings which should not significantly affect present health, but may adversely affect future health
4	Below average health — has present health problem which does not significanly affect ability to perform present job but will affect long-range prognosis
5	Poor health — has significant health problems at present time which will probably affect ability to perform present job or may cause significant absenteeism in near future
6	Unemployable because of severe present illness or physical defect

in which candidates had been classified according to Sawyer's system. He found that 20% of the applicants were graded as A, 58% as B, 10% as C and 12% as D.

A slightly simpler classification system recommended by the American Medical Association (AMA, 1973) is shown in Table 2.3.

A very comprehensive classification for the traditional fitness-for-work examination was developed by Lincoln (1968) for his work with the US Atomic Energy Commission. He categorized his candidates with respect to their physical health, their mental health, and their overall suitability, using in each case five or more different levels of fitness. His criteria for physical health are given in Table 2.4 and his criteria for mental health are given in Table 2.5.

It will be noted, of course, that Lincoln's (1968) criteria for mental health appear to be more concerned with the definition of motivation and career potential than with the state of the applicant's health, and indeed his classification might very well not be

Table 2.5. Criteria for mental health as suggested by Lincoln (1968)

Class	Description
1	Outstanding candidate — exceptional motivation; life and career adjustment far above average; should be expected to realize potential and maintain high level of productivity within training and ability
2	Above average candidate — can be expected to make above average contributions; no major mental health problems in past history and none anticipated; career and life adjustment in good balance
3	Average candidate — no major problems at present time but motivation or life and career adjustment not outstanding
4	Marginal candidate — has had significant career or personal problems which will probably prevent applicant from realizing his potential and may create future personnel problems
5	Poor risk candidate — history of poor career or personal adjustment; may have had psychiatric illness with recovery; likelihood of future problem is high
6	Unemployable — history of mental illness with evidence of severe difficulty at present time

Table 2.6. Overall suitability of candidate for job according to Lincoln (1968)

Class	Description
1	Top candidate — 'ball of fire'
2	Excellent prospect
3	Average candidate
4	Poor risk — don't employ unless no other candidate available
5	Unemployable

found acceptable either to human resources personnel or to human rights advocates.

Using both sets of criteria, he defines the overall suitability of a candidate as given in Table 2.6

It will be shown later that there are more appropriate methods for classifying a candidate for employment in the light of the job demands and the candidate's functional capacity, but for use with a traditional fitness-for-work examination Lincoln's criteria for physical health would appear to provide one of the most comprehensive and useful classification systems.

Work restrictions

On completion of a medical examination in which an applicant is found to have some temporary or permanent degree of incapacitation, it is common for the examiner not only to define the health status by letter or number, but also to assign some temporary or permanent restriction on the nature of the duties that the applicant may be permitted to perform, or to reject the applicant as being unemployable. Partly because the examination is oriented towards functional incapacity rather than functional capacity, and partly because the examiner may not be aware of the task demands, a rejection may in fact be unwarranted, and any restriction applied may be vague, not very useful to the employer, and may be couched in terms such as 'fit for light work only', 'no heavy lifting', and so on. A more specific approach to the application of restrictions will be discussed later in consideration of job matching.

Causes for rejection

It is the normal practice for the examiner in the traditional pre-employment examination to use his/her best clinical judgment in determining whether or not an applicant should be rejected for employment. This judgment naturally varies between individuals and is tempered by an appreciation of the likely prognosis of any functional incapacity and an understanding of the needs of the workplace, as well as by the increasing demands of human rights legislation. The previously noted AMA *Guidelines* (1973) outline some very simple criteria to assist in making this judgment and state that the only bars to immediate employment in non-hazardous occupations should be:

(a) communicable disease,
(b) incapacitating injury or disease,
(c) mental illness in which impaired judgment or actions prevent co-operative effort.

While the above criteria are simple in principle, unfortunately the term incapacitating is not defined and the examiner is again left to the resources of his/her own judgment.

In general, however, as indicated in the discussion of classification scales, applicants are commonly rejected if they are unable to perform the assigned job without being a hazard to themselves or others. This situation may arise by reason of chronic illness not amenable to treatment, physical defect from birth, illness or injury rendering them incapable of work, disease which can be easily communicated by normal association, inability to communicate, or major disfigurement.

In this connection Campione (1972) made a study of pre-employment examinations for all applicants for employment in a corporate office during 1950–70. In addition to what might be expected to be the more justifiable clinical causes for rejection, he found that some of the more common and recurring factors primarily responsible for the rejection of applicants were:

(a) active disease,
(b) two or more moderate physical impairments,
(c) concealment of previous medical attention,
(d) multiplicity of small complaints,
(e) symptoms attributed to previous employment,
(f) emotional maladjustment,
(g) adverse corporate experience of similar disease,
(h) serious disease with poor prognosis,
(i) pregnancy,
(j) resentful attitude of the applicant to evaluation.

While attitudes towards human rights and expectations in the workplace have changed considerably since that time, it is likely that many of the same reasons can still be found in various companies.

Records and disposition of findings

It is not only necessary, but it is mandated that written records of all medical findings be maintained. These records are confidential and must be maintained in secure storage. Where a medical department exists the records are kept within it. Where there is no medical department they are normally kept within a personnel department, but separate from other personnel records. It is emphasized that, as in any medical examination, the information content of an examination for fitness or work is confidential. While all significant findings should be discussed with the applicant with a view to emphasizing the importance of seeking medical care where necessary, information may not be passed on to any other person without the expressed request or consent (preferably in writing) of the applicant. The AMA *Guidelines* (1973) point out that with the consent of the applicant a transcript of the findings, or pertinent data, may be submitted to another physician or to a health agency, or for example as required for insurance purposes. The employer, however, should be provided only with a classification of the applicant's fitness, with or without associated restrictions and recommendations. A specialized form is often used for this purpose. In additior, they note that government agencies such as courts, occupational safety and health agencies, workers' compensation agencies, or other health authorities may be supplied with information as required by law.

Effectiveness and validity of pre-placement examinations

Many, if not most, medical practitioners have questioned the effectiveness and validity of routine pre-placement examinations, but have dutifully continued to conduct them for want of a suitable alternative. More than 25 years ago, Thrift Hanks, then medical director of the Boeing Corporation, stated:

> The routine physical examination has serious shortcomings even if supplemented to absurd lengths by . . . laboratory and x-ray procedures of little value and some potential hazard. It is at best a gross screening device whose chief findings can be readily duplicated by the use of a health questionnaire or a trained nurse-interviewer. Ultimate placement of the good worker with easy adaptability is no problem. Physical disability is seldom the point at issue. Lack of skill, lack of adaptability, poor work record, inability to accept responsibility, poor relationships with fellow workers and supervisors, and an excessive concern with self are much more important.
>
> Knowledge of a specific disability discovered by physical examination is not reliably preventative of injury and is not preventative at all of claims for workmen's compensation. Conversely the failure to find disability protects neither the worker nor the employer. (Hanks, 1962)

It is doubtful if conditions have changed much, if at all since that time. Indeed, nearly 15 years later Alexander and his colleagues (1975) conducted a study comparing attendance and work performance of job applicants identified by pre-placement examination to be at increased risk with a normal risk group over a period of three months after hire. Data were based on the findings from 5936 male and female applicants over 18 months.

For the control group the actual medical recommendation was given regarding the anticipated work performance or attendance, identifying health problems or work restrictions. For the trial group the information given to the employment office was as if the applicants were medically acceptable without restrictions.

During the first 3 months after hiring no significant differences could be identified with respect to job performance, supervisor's recommendation for continued hiring, capability of doing the tasks demanded and personal (non-medical) absence. The evaluation on the other hand did serve a purpose in predicting sickness absence during the 3 months.

Like others who have considered the usefulness of pre-placement examinations, such as Harte (1974) and Schussler and his colleagues (1975), whose work has been considered earlier in connection with establishing the usefulness of questionnaires and nurse interviews, Alexander *et al.* (1975) note that there is no disagreement on the value of pre-employment evaluations for applicants for hazardous jobs. However, they also point out that using that type of evaluation simply to screen out generally undesirable applicants neither benefits the applicant nor provides medical records which could be considered accurate or reliable. As noted in their study, analysis of the three month supervisory appraisals does not support the contention that poor work performance may be predicted by a pre-employment medical evaluation. Work-force losses during the first three months after hire also showed no difference between the apparent 'no risk' group and 'risk' group. As a conclusion they state categorically that '. . . there is serious doubt that a cost benefit from a pre-employment health evaluation for non-hazardous jobs is realized by screening out attendance risks.'

Before discussing any alternative to the routine pre-placement examination it is desirable to examine the traditional role of management in the employee selection process. This will be considered in the next two chapters.

Chapter 3
Job evaluation and analysis

The selection and hiring of employees and the evaluation of jobs is of course a management function. The process has evolved from a simple interview of a prospective candidate by a foreman or line manager to a complex procedure conducted by human resources personnel and their specialist advisors and assistants. The simple interview attempts in the subjective judgment of the interviewer to determine the applicant's suitability to do the task assigned to him. In small companies, where the jobs are few and tasks well known, and where even the candidate may be well known to the interviewer, it may serve a valuable purpose. In large companies, where jobs are many and tasks complex, the simple interview has to be replaced by a multivariate procedure which attempts to ensure that the applicant meets all the prescribed requirements for hiring.

Thus, the procedure of job evaluation and analysis has been used for several purposes. Primarily it has been used to categorize jobs in terms of their relative financial worth, that is, to determine relative pay scales. It has also been used by those concerned with various forms of time and motion study to determine the time taken to perform various tasks and their component subtasks, again partly to determine wage-incentive plans, but also partly to improve the balance of production lines and the work of crews of workers. More recently, however, and this will be discussed at length in a later chapter, job evaluation and analysis has been used in an attempt to match the demands of the job to the capacities of the worker.

The concept of job evaluation is not new. According to Patton *et al.* (1964), the US Civil Service Commission made an initial, if limited, attempt to do so in 1871. More lasting approaches were introduced much later by Lott (1926) in the USA, (quoted by Livy, 1975) although it was not until World War II, again in the USA, that the concept began to become commonplace there, spreading to Britain, Europe and elsewhere many years after.

A comprehensive study of the nature and approaches to job evaluation and analysis was undertaken by McCormick (1979), who was also widely known for his definitive texts in human factors engineering. Much of the material in this chapter is derived from McCormick and from comprehensive studies such as those presented by Bartley (1981), Fine (1974), Gael (1983), and Livy (1975), as well as others referenced in the text. It is emphasized, however, that the primary orientation of most of these works is not so much to match the work capacities of the applicant to the work demands of the job, as to determine the intrinsic requirements of the job or tasks in question for wage and production purposes.

Definitions

The following definitions are derived from McCormick (1979), utilizing information from the *Handbook for Analyzing Jobs* (US Department of Labor, 1972):

Occupation: A system of jobs of a general class, on an across-the-board basis, without regard to organizational lines.

Job: Group of tasks which are identical with respect to their major or significant functions and sufficiently alike to justify their being covered by a single analysis. There may be one or many persons employed on the same job.

Position: Complex of tasks and duties for any individual. A position exists whether occupied or not.

Duty: Used rather loosely to refer to a large segment of work performed by an individual. Represents one of the distinct major activities involved in the work performed, and consists of several tasks that are or may be related.

Task: A discrete unit of work performed by an individual. Usually comprises a logical and necessary step in the performance of a duty, and typically has an identifiable beginning and end.

Task element: The smallest step into which it is practical to subdivide any work activity without analysing the separate motions, movements and mental processes involved.

Elemental motion: Very specific separate motions or movements, as used in time-methods analysis procedures in industrial engineering.

Function (Gael, 1983): Broad subdivision of a job comprising a group of tasks that are somewhat related because of the nature of the work or the behaviour involved.

Gael (1983) goes on to add that there are two types of function, firstly, supervisory, as in organizing, planning, directing, developing and so on, and secondly, direct work as in maintaining, repairing, operating and so on. He considers that a job is an amalgam of functions performed by individual employees.

One of the major characteristics of a task is that it is directed towards an objective, a common goal, or outcome, and that it is an orderly, homogeneous grouping of goal-oriented human activities applied methodically to things or equipment and usually performed by one person in less than a day (Terry and Evans, 1973). The last comment introduces an element of time into the definition which may or may not be applicable.

Job analysis (Livy, 1975): a process of investigation into the activities of work and the demands made on the workers, irrespective of the type or level of employment.

Job description, job specification (Livy, 1975): Two related terms sometimes used indiscriminately. Job description is best regarded as a written statement of the main duties and responsibilities which a job entails, and job specification as a list of criteria in terms of the personal capacities and inclinations deemed to be necessary for successful job performance.

Qualitative approaches to job evaluation

Livy (1975) has written a comprehensive review of the principles of job evaluation. He discusses a number of different techniques from the viewpoint of establishing wage parities and pay ratings, noting that both non-quantitative and quantitative methods can be applied. Of the non-quantitative methods the two most common are ranking and classification.

Ranking

Ranking provides the easiest approach. It does not call for much detail and can be executed relatively quickly with a minimum of time, energy, and resources. Jobs are not broken down to their component elements, as occurs in other systems, but are compared as total entities. The method however is non-quantitative in that it does not give any indication of the degree of difference between jobs, only that one job is more or less demanding of the individual, or important to the organization, than another.

It begins with a qualitative analysis. Information about the job is assembled and recorded in a standard format. It is important in this regard to rank the job and not the incumbent. For ranking purposes it is desirable to define a key or bench-mark job or jobs which can be used as a yardstick against which other jobs can be compared. Preferably bench-mark jobs should be those which represent a wide range of job requirements and should be selected at various levels in the organization. Other jobs are then compared by determining whether the job in question is more or less important than the bench-mark job. However, when a partial hierarchical outline beings to emerge, ranking requires as much comparison to be made against jobs, which have already been placed as it does against bench-mark jobs. Common practice is to reduce the job analysis information or job descriptions to specially prepared cards. Cards can then be moved into the appropriate position as the jobs demand.

Although quick and easy, the method cannot be considered accurate or precise; it does not, indeed, define a scientifically discriminating series of jobs in hierarchical order. If greater degrees of precision are required then some other technique is needed. A more precise approach, although still not quantitative, is given by the process of classification.

Classification

The classification method, sometimes known as the grade description system, or predetermined ratings, is a ranking operation in that it ends with jobs being organized into collective grades, but the procedure for achieving these is reversed in comparison with the ranking method. A good brief review of various types of classification systems is presented by Frank (1982).

The starting point is to define the form and size of the organizational structure desired, for example, flexible, capable of relatively easy dynamic change as a business organization, or rigid and uniform as in a government division. The system lends itself most easily to a fairly rigid organization as in a civil service, and indeed it originated in the US Civil Service by the 1949 Classification Act which established a General Schedule of 18 grades covering most professional, administrative and technical jobs (Livy, 1975). These ranged from GS-1, concerning simple routine jobs through to GS-18, covering exceptionally demanding difficult jobs. In general, jobs from GS-1 to GS-4 require less than college or university education; jobs in GS-5 to GS-11 are medium level technical, administrative and supervisory positions, and jobs above GS-11 are distinctly high-level jobs, with compensation rates reflecting substantial differences between grades. Similar system are found among civil services in many other countries.

Having defined the desired structural organization, the next step is to establish an arbitrary but reasonable number of job grades, providing for administrative purposes a series of occupational levels broadly distinguishable from each other in terms of their activities and needs. These should reflect activities as defined in the conventional organization chart. The responsibilities within each of these grades must then be clearly defined in unequivocal terms, using bench-marks as required.

The system, of course, is somewhat arbitrary. Livy (1975) points out that criticism has been levelled mainly on the grounds of lack of objectivity in assessing a job before allocating it to a particular grade. Mistakes and misconceptions can undermine the accuracy of the grading and, in turn, diminish the confidence of the employees whose careers may be affected by it. Consequently, more definitive systems have been demanded, such as those involving analysis of individual jobs.

In his comprehensive review of job evaluation and job analysis, McCormick (1979) defines three sets of approaches to job analysis, namely, conventional job analysis, methods and task analysis and structured job analysis. Methods and task analysis is a set of industrial engineering techniques concerned largely with time and motion study and will not be considered further here.

Conventional job analysis programmes involve the assembly of job-related information by observation of, and/or interview with, job incumbents, and the preparation of job descriptions in a standard written format. While various modifications have been made, these all derive originally from the programme developed by the US Employment Service of the Training and Employment Administration (USES) as presented in the *Handbook for Analyzing Jobs* (US Department of Labor, 1972). Since the USES approach is the basis of many other evaluation systems and is significant in the development of the approach to physical demands analysis, it is enlarged upon in the next section.

USES job analysis

The USES job analysis record is presented on a standardized form outlining job information in predetermined standardized terms which are specified in *Handbook for Analyzing Jobs* (US Department of Labor, 1972). The concept of the analysis is based on defining the extent of involvement of the worker in three areas, namely, data, people and things. Functions within each of these three areas are defined in Table 3.1

Each function is clearly defined in the *Handbook* and is also number coded. In practice, the job is first carefully defined by observation and interview with the incumbent and his/her superiors and entered by title on the form. A concise summary of the work of the job is added in the appropriate box along with the highest level of

Table 3.1. USES worker functions (after US Department of Labor, 1972).

Data	People	Things
0 Synthesizing	0 Mentoring	0 Setting-up
1 Co-ordinating	1 Negotiating	1 Precision working
2 Analysing	2 Instructing	2 Operating/controlling
3 Compiling	3 Supervising	3 Driving/operating
4 Computing	4 Diverting	4 Manipulating
5 Copying	5 Persuading	5 Tending
6 Comparing	6 Speaking, signalling	6 Fending, offbearing
	7 Serving	7 Handling
	8 Taking instructions, helping	

Table 3.2. Examples of work functions (after McCormick, 1979, from USES Handbook for Analysing Jobs).

— Shovels coal into mine cars for haulage (things relationship, non-machine, handling)
— Examines structural aircraft assemblies to verify conformance to specifications (data and things relationship, analysing and handling levels)
— Solves problems in mathematics in such fields as engineering, physics and astronomy (data relationship, synthesizing level)
— Supervises and co-ordinates activities of carpenters on housebuilding project (data, people and things relationship, co-ordinating, supervising and precision working levels)

function in terms of the previously noted data, people and things. Examples of differing work functions from the USES *Handbook* are given in Table 3.2.

The aids and equipment used by the worker are also defined as part of the work of the job, under the heading of MPSMS (materials, products, subject matter and services). An example of a workfield from the *Handbook* is given in Table 3.3.

To assist in the analytical procedure, worker traits that are considered important to the job are also recorded in standardized terms. These are defined as general educational development (GED), specific vocational preparation (SVP), aptitudes, interests, temperaments, physical capacities and adaptability to environmental conditions.

The GED scale comprises three divisions: reasoning development, mathematical development and language development, each of which is further divided into six levels, the requirements for which are outlined in the *Handbook*.

Aptitudes are defined as the specific abilities required of a worker in order to facilitate the learning of some task or duty. Eleven of these are defined, each of which is then rated on a defined scale of significance. The following is a brief interpretation derived from the *Handbook*:

Intelligence: General learning ability; ability to understand instructions and principles.
Verbal: Ability to understand meanings of works and ideas, and to use them effectively.
Numerical: Ability to perform mathematical operations.
Spatial: Ability to comprehend forms in space.

Table 3.3. Examples of USES work field (after McCormick, 1979, from USES Handbook for Analysing Jobs).

COOKING — FOOD PREPARATION

Preparing foods for human or animal consumption, by way of any combination of methods which may include methods specific to other work fields such as *Baking-drying, mixing, shearing-shaving, stock checking, and weighing.*

Methods verbs

Basting	Curing	Measuring	Roasting
Broiling	Flavoring	Pasteurizing	Rolling
Brewing	Frying	Pickling	Seasoning
Churning	Heating	Rendering	Spreading
	Kneading		

Machines	*Tools*	*Equipment*	*Work aids*
Continuous churn	Cleaver	Broilers	Charts
Pasteurizer	Cutters	Grills	Dishes
Vane churn	Forks	Ovens	Hoppers
	Ice picks	Ranges	Kettles
	Knives	Roasters	Mixing bowls
	Paddles	Smoke chambers	Pans
	Sifters	Steam digesters	Pots
	Spatulas		Recipes
	Spoons		Storage bins
			Storage tanks

Controls battery of smoke chambers in which such meat products as bacon, ham, meat loaf, sausage, shoulders, and weiners are cooked and cured

Mixes and bakes ingredients, according to recipes, to produce breads, pastries, and other baked goods

Mixes, cooks, and freezes ingredients to prepare frozen desserts such as sherbets, ice cream, and custards

Operates ovens to roast dry breakfast cereals made from corn, rice, bran, and oats

Plans menus and cooks meals in private home, according to recipes or tastes of employer

Form perception: Ability to perceive pertinent detail in objects or in pictorial or graphic material.
Clerical perception: Ability to perceive pertinent detail in verbal or tabular material.
Motor co-ordination: Ability to co-ordinate hands and eyes or fingers rapidly and accurately.
Finger dexterity: Ability to move fingers and manipulate small objects.
Manual dexterity: Ability to move hands easily and skilfully.
Eye-hand-foot co-ordination: Ability to move hand and foot co-ordinately with each other in accordance with visual stimuli.
Colour discrimination: Ability to perceive or recognize similarities or differences in colours, or shades or values of the same colour.

In addition to aptitudes there is provision for definition of 10 temperament factors which refer not so much to the characteristics of the person as to the different types of occupational situations to which the worker must adjust. For instance, one such defined situation involves a variety of duties often characterized by frequent change; another involves doing things only under specific instruction with little or no provision for independent action, and so on.

While much of the process is concerned with describing the nature of the job, provision is also made for description of the physical demands of the job in terms of six different factors. The first of these is strength, and specifically strength required for lifting, carrying, pushing and pulling. Strength is graduated in five required levels: sedentary, light, medium, heavy and very heavy, each of which is further defined in terms of maximum weight of lift and requirements for carrying and so on. The second factor refers to the need for climbing and/or balancing, and is defined quantitatively in terms of, for example, the use of stairs, scaffolding and so on, and the requirements for maintaining body equilibrium under difficult circumstances.

The third factor defines the need for stooping, kneeling, crouching and/or crawling, again in qualitative terms, while the same approach applies to the fourth factor which defines the needs for reaching, handling, fingering and/or feeling. The fifth factor describes the need for talking and hearing, while the sixth describes that for seeing. Seeing is divided still further into requirements for near and far acuity, and colour vision.

While the response to some of these items is commonly given in percentage, weight or even decibels, many can be coded as follows:

NP: not present (does not exist),
O: occasional (exists up to $\frac{1}{3}$ of the time),
F: frequent (exists from $\frac{1}{2}$ to $\frac{2}{3}$ of the time),
C: constant (exists $\frac{2}{3}$ or more of the time).

A description of the working environment is included in the schedule, subdivided into seven factors: (a) work location, that is, protection from weather but not necessarily temperature change, (b) extremes of heat with temperature variation, (c) extremes of cold with temperature variation, (d) wetness and humidity, (e) noise and vibration, (f) hazards, that is, risk of bodily injury, (g) fumes, odour, toxic conditions, dust and poor ventilation.

The USES form also includes, as an integral part, demographic and personal data pertaining to the incumbent employee, such as general education and vocational perparation, in-house and on-the-job training, experience, apprenticeship (where applicable), along with information pertinent to his/her career development at work. The form in its totality thus gives a comprehensive qualitative view of the job and the employee at hand.

Functional job analysis

The USES approach to job analysis is effective and comprehensive. Since, however, the formulation of the descriptions is to a considerable extent left to the skill and discretion of the analyst, the descriptions and end results may not be consistent

Table 3.4. Worker function scales for determination of Functional Job Analysis (Fine, 1974).

Data	People	Things
Synthesizing	Mentoring	Precision working Setting-up
Co-ordinating Innovating	Negotiating	Manipulating Operating–controlling Driving–controlling
Analysing Computing Compiling	Supervising Consulting, instructing treating Coaching, persuading diverting	Handling Feeding–offbearing Tending
Copying	Exchanging information	
Comparing	Taking instructions helping, serving	

among analysts, or even for the same analyst, at different times. This and other problems were addressed by Fine and his colleagues initially in developing the 1965 edition of the *US Dictionary of Occupational Titles* for the USES, and led to the development of a more sophisticated system known as functional job analysis (FJA). The results of their research are presented in a number of papers (Fine, 1973, 1974; Fine *et al.*, 1974; Fine and Wiley, 1971). It might also be noted that the practice of the USES system consists of assigning levels to jobs, whereas Fine and Wiley (1971) consider the application to individual tasks, and, by summation of tasks, ultimately to the entire job.

Fine *et al.* (1974) note that in the process of FJA a basic distinction is made between what workers do and what gets done — between behaviour and end results. This distinction is carried into the methods of analysis, that is, the gathering of data, and the formulation of the task statements, since historically most job descriptions dwell primarily on what gets done. In FJA task statements are verbal formulations of activities that make it possible to describe what workers do and what gets done so that recruitment, selection, assignment, training, performance evaluation and payment can be efficiently and equitably carried out. The focus of the attention, then, must be on the words and the organization of words used in the task statement to express the task. The resulting statement must be as close to reality as possible, and be expressed in such a manner that the language of one task statement is compatible with that of others in that context, and that together they can describe the technology of a work situation.

Thus, a task in FJA is defined in terms of a controlled language, a controlled method of formulation, and in relation to a systems context, thus:

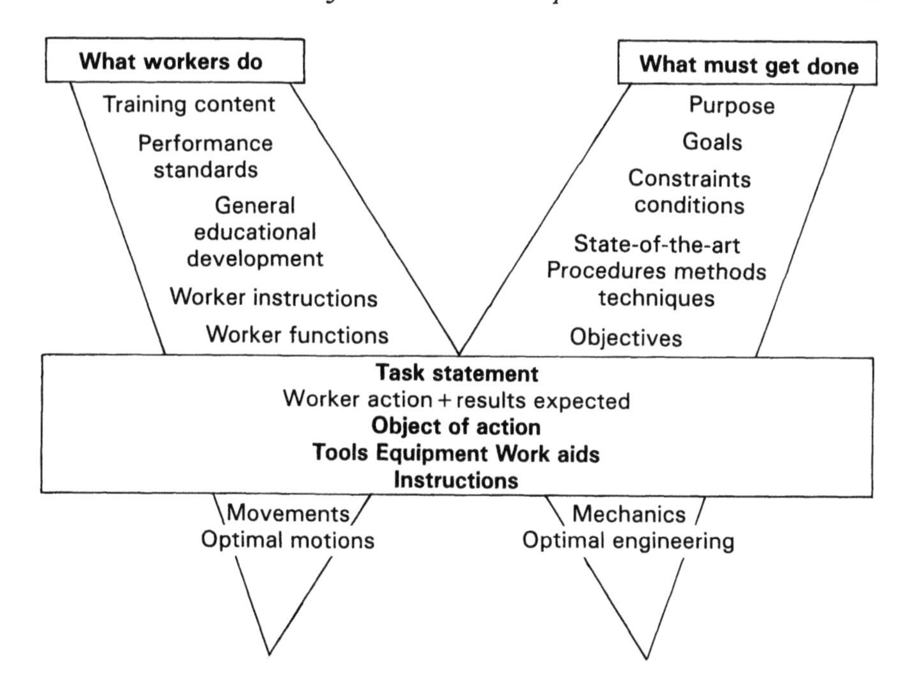

Figure 3.1. Model of task statement (after Fine, 1974)

> A task is an action sequence grouped through time designed to contribute a specified end-result to the accomplishment of an objective and for which functional levels and orientation can be reliably assigned. The task action or action sequence may be primarily physical, such as operating a typewriter, primarily mental, such as analysing data, and/or primarily interpersonal such as consulting with another person. (Fine *et al.*, 1974)

As with the standard USES approach, FJA utilizes the concept of data, people and things. Table 3.4 (Fine, 1974) represents an analysis of what workers do in their work assignments.

The 28 functions shown in Table 3.4 were primarily generated by analysing the language used to describe what workers did and what got done in 4000 job definitions of the second edition of the *Directionary of Occupational Titles*. In that study it was shown that while some 2000 such verbs were used few of these actually described worker behaviour and, indeed, those noted in Table 3.4 above actually defined virtually all the actions. Fine (1974) notes by selecting one verb from each of the three scales it is possible to describe any task, and, in summary form, the job performed by any worker.

Figure 3.1 (Fine, 1974) is a model of a task statement in functional job analysis and is an explicit expression of a worker action and the expected result of that action.

To make the statement meaningful and self-sufficient the action verb is modified by the means used (tools, methods and equipment), by the immediate object of the action if different from the result, and by some indication of the prescription/discretion in the worker instruction. Table 3.5 (Fine, 1974) is an

Table 3.5 Example of functional job analysis (after Fine, 1974).

(a) TO DO *THIS* TASK

Data	People	Things	Data	People	Things		Reas.	Math.	Lang.	
W.F. level			W.F. orientation			INSTR	G.E.O.			Task No.
3B	1A	2B	70%	5%	20%	2	3	1	4	

Goal: To be completed by individual Objective: To be completed by individual

Task: Types/transcribes standard form letter, including specified information from records provided, following S.O.P. for form letter, but adopting standard form as required for clarity and smoothness, etc., in order to prepare letter for mailing.

(b) TO THESE STANDARDS

Performance standards
Descriptive
  Type with reasonable speed and accuracy
  Format of letter is correct
  Any changes/adjustments are made correctly

Numerical
  Completes letter in % period of time
  No uncorrected typing, mechanical or adjustment errors per letter
  Fewer than % omissions of information per % number letters typed

(c) THE WORKER NEEDS *THIS* TRAINING

Training content
Functional
  How to type letters
  How to transcribe material, correcting mechanical errors
  How to combine two written sets of data into one
Specific
  How to obtain records and find information in them
  Knowledge of S.O.P. for standard letter format: how/where to include information
  Knowledge of information required in letter
  How to use particular typewriter provided

example of what the master tool looks like with the applications paradigm (TO DO THIS TASK — TO THESE STANDARDS — THE WORKER NEEDS THIS TRAINING) superimposed.

In this connection, Fine draws attention to the section labelled TO DO THIS TASK, noting that it is written according to a definite structure which is controlled for content and reliability, the controls being represented by the 10 figures shown at the top. Thus, for the task shown, the ratings show that is a task involving compiling, taking instructions and operating-controlling, where the orientation, that is the relative involvement with data, people and things, is 70%, 5%, and 25%, respectively. The remainder of the ratings relate to four additional scales contained in the USES *Handbook* (US Department of Labor, 1972), namely scales of instruction, reasoning, mathematics, and language, in which the levels of rating are functional rather than academic.

The standards so indicated act as a basis for the management to achieve productive work within the organization, and as a guide to the worker in terms of what is expected of him.

As implied above, in so far as their job context is concerned, workers in relation to things draw upon physical resources; in relation to data they draw upon mental resources, thus, only a few definitive functions are involved. With respect to machines, workers act to feed, tend, operate or set them up; with respect to vehicles or similar devices they act to drive/control them.

Just as the functions in the USES system are defined in hierarchical terms under the headings of data, people and things (e.g. synthesizing, co-ordinating, analysing, computing and so on, with respect to data), similar hierarchies are organized in FJA, with the one difference that the numbering system is reversed. Thus, the first function under data for USES is listed as zero despite the fact that it is one of the highest order; the same function in FJA is listed as 1. Similar changes occur for the hierarchies of people and things.

The FJA approach also provides for analysis in terms of worker function scales, namely, the level and the orientation of involvement with the three hierarchies. The level is similar to that for USES; the orientation reflects the relative involvement of the worker at that level and is expressed by assigning a percentage in increments of 5%, as shown in Table 3.6 (Fine and Wiley, 1971).

The worker function scales are defined in very much greater detail than in the USES system, although, as noted, the headings remain the same. In addition, some of the headings are further subdivided into additional levels, A,B, and C. For example, under the heading of Data Function Scale the acts of (1) comparing and (5B) co-ordinating are as follows:

1. *Comparing*: selects, sorts, or arranges data, people or things, judging whether their readily observable functional, structural, or compositional characteristics are similar to or different from prescribed standards.

5B. *Co-ordinating*: decides time, place, and sequence of operations of a process, system, or organization, and/or the need for revision of goals, policies (boundary conditions), or procedures on the basis of analysis of data and of performance review of pertinent objectives and requirements. Includes overseeing and/or executing decisions and/or reporting on events.

As already noted, the FJA system also provides for the development of four descriptive scales (Fine, 1973) in a similar manner to those for the USES system, namely, worker instructions, reasoning development, mathematical development and language development.

Throughout the system emphasis is placed on the usage of standardized, consistent work statements, using standardized and defined words, beginning with some action verb, modified by the means, such as tools, methods and equipment which are required by the immediate objective of the action, and by some indication of what the worker must or may do to complete the action.

Table 3.6. Example of worker function scales (after Fine and Wiley, 1971).

Task

Asks client questions, listens to responses, writes answers on standard form, exercising leeway as to sequence of questions in order to record basic identifying information.

Analysis of task

Area	Functional level	Orientation
Data	Copying	50%
People	Exchanging information	40%
Things	Handling	10%

The foregoing techniques are examples of qualitative techniques of job evaluation which, as already noted, have some bearing on the development of what will later be examined as functional capacity evaluation and physical demands analysis. Other forms of so-called quantitative approach have also been developed, some of which are examined here.

Quantitative approaches to job evaluation

The use of the term quantitative in describing techniques of job evaluation, although common, is slightly misleading. In general it refers to the application of numbers to subjectively or quasi-objectively derived evaluations or analysis in such a manner that one can be compared more readily against another. An early example of quantitative evaluation is that of points rating.

Points rating

The points rating approach was first developed by Lott (1926) for use in the Sperry Gyroscope Company, Inc. Since then it has undergone considerable change, but in its current form is the most widely used system today. The method is both analytical and quantitative in that jobs are broken down into component parts for comparison purposes, following which each element is assigned a numerical rating. The total value of the numerical points in a job constitutes the rating of that job.

The process begins by defining criteria against which jobs and/or job elements are compared. The criteria are derived from consideration of the most essential elements common to the range of jobs to be evaluated, and are commonly known as job factors. Where large numbers of jobs are involved it may become cumbersome and ineffective to define and use criteria applicable to the entire gamut of occupations. In such cases it may be necessary to define families of groups of jobs, such as manual, clerical, managerial and so on, each with its own points rating system, but each consistent with the overall plan. Where some characteristics are critical to a job, it is often desirable also to attach some specific weighting factor or factors.

The main steps in the process, according to Livy (1975) are:

1. Establish a representative committee with responsibility for the job evaluation.
2. Analyse a significant sample of jobs and prepare job descriptions and specifications (or job information sheets).
3. Select and define those factors which are considered to be the most critical in determining the relative degrees of difficulty and responsibility between jobs. Too few factors will reduce the discriminatory power of the technique, and too many will introduce problems of co-variance.
4. Weight the factors according to their relative importance. Commonly this entails dividing each factor into a number of degrees representing the extent to which a given factor may be present in a job, and awarding point values pertinent to each degree.
5. Test a selected number of key jobs to determine if the dispersion in the relative rating of the jobs conforms with the existing differentiation between jobs, or with the results obtained by some other method, cross-checking factor by factor to ensure that there are no obvious discrepancies and that reasonable reliability exists.

An example is shown in the points rating system devised by Kress (1939) for the US National Electrical Manufacturers Assocation (NEMA) in which he defined 11 characteristics for the jobs under consideration, each characteristic being subdivided into a number of specific subfactors, as shown in Table 3.7.

Table 3.7. NEMA job analysis system (Kress, 1939).

Generic factors	Specific subfactors
Skill	Experience
	Education
	Initiative and ingenuity
Effort	Physical demand
	Mental and/or visual demand
Responsibility	For equipment or process
	For materials or product
	For safety of others
	For work of others
Job conditions	Working conditions
	Hazards

In this scheme each subfactor can be further divided into degrees and weighted according to its importance with respect to other factors (see Table 3.8). The use of degrees indicates the extent to which a subfactor may be present in a given job. Any number of degrees can be used; a five-point scale is merely an administrative convenience. Once the degrees have been defined, points can be attached to each degree. For example, as illustrated in Table 3.8, points might be assigned to degrees of experience beginning with 20 and extending to 100. Again, however, the assignment can be arbitrary, although it must be consistent with the overall scheme. Very sophisticated programmes can be developed to meet many different requirements.

Table 3.8. Example of degrees of experience for use in points rating system (Livy, 1975).

Subfactor degree	Experience	Points awarded
1st degree	Up to 3 months	20
2nd degree	3 months–1 year	30
3rd degree	1–3 years	45
4th degree	3–5 years	60
5th degree	Over 5 years	100

Task inventories

The task inventory is a form of job evaluation questionnaire comprising a listing of tasks within some occupational field, along with an indication of their occurrence in the work of the incumbent and the relative amount of time spent on each. The technique was developed by Morsh and Christabel (1966) in the US Air Force. Melching and Borcher (1973) note that there are generally two types of task inventory, namely, that concerned with supervisory activities such as supervising, implementing, training, inspecting and evaluating, and that concerned with work performance, such as performing, maintaining, troubleshooting, repairing, removing and replacing, adjusting and installing.

Each inventory is recorded on a form devised for the purpose which includes space for listing the tasks, and columns for checking the significance to the incumbent and the time spent. The inventory indicates what is done in the job, but not how or why.

The response scales used for recording the evaluation define primary and secondary rating factors (Morsh and Archer, 1967). The primary rating scales define the extent to which an incumbent is involved in a particular task, thus:

— Importance (i.e to the job)
— Part-of-job scale (i.e extent to which task is part of position)
  0. not part of job
  1. under unusual circumstances may be minor part of position
  2.
  3.
  4. substantial part of position
  5.
  6. significant part of position
— Performance (whether incumbent performs task or not)
— Frequency of performance (time per day, per week, etc.)
— Time spent (actual time spent on task)
— Relative time spent (time spent relative to total time)
  A relative time scale is as follows:
  1. very much below average
  2. below average
  3. slightly below average
  4. about average
  5. slightly above average

6. above average
7. very much above average.

The secondary rating factors provide information, usually, about the tasks themselves. McCormick (1979) lists some of those that have been used, as follows:

— complexity of task
— criticality of task
— difficulty of learning task
— where incumbent learned task (training course, on-the-job, etc.)
— where respondent believes task should be learned
— special training required
— time considered necessary to learn task (broad terms, e.g. a few weeks)
— difficulty in performance
— technical assistance required in performance
— supervision required in performance
— satisfaction derived in performance.

Secondary responses can be completed by the incumbent, but are more commonly completed by a supervisor or other expert. The responses are rated according to a 7- or 10-point scale.

In developing a task inventory the first step is to define the scope of the inventory, that is, the breadth of the occupational area to be covered, and the second step is to locate and acquire source materials pertaining to the area defined. These might include training materials, instructional materials and manuals/textbooks, as well as previously constructed task inventories or other forms of job analyses.

The third step is development of the inventory itself. There are two approaches to this: (a) develop a stock of task statements or descriptions from all sources and organize them into appropriate groups of duties, and (b) define a general outline of duties and develop appropriate task statements within each duty or subdivision of a duty. In this connection tasks can be considered as job activities that can be identified in terms of discrete units of work. Care should be taken to avoid excessive specificity and excessive generality.

Guidelines for the preparation of task statements have been developed by Melching and Borcher (1973), as follows:

— The task statement must be so clear that it is easily understood by the worker.
— The task statement must be stated using terminology that is consistent with current usage in the occupational area.
— The task statement should be brief to save reading time of the worker.
— The task statement must be written clearly so that it has the same meaning for all workers in the occupational area.
— Abbreviations must be used cautiously since they may not be understood throughout the whole occupational area.
— The task statement must be worded so that the task rating scales make good sense when applied to it.
— The task statement must be rateable in terms of time spent and other rating factors — eliminate such phrases as 'have responsibility for', 'understand', 'have knowledge of'.
— Vague or ambiguous words should be avoided.

— Short words should be used in preference to long words.
— The task statement should begin with an action word in the present tense, with subject 'I' understood.
— The task statements should be arranged alphabetically under each duty to allow incumbent to scan list and readily identify required task.

After the inventory is complete it should be reviewed by experts for clarity, for errors and omissions, and for modifications and re-groupings, if any, preparatory to being used in a pilot test to determine its suitability. After final revision it is ready for use.

McCormick (1979) has examined the value of task inventories as a means of job analysis. He states:

> In this regard, although the reliability of responses of individuals to individual tasks tends to be rather moderate, when the many bits and pieces of inventory data are pooled, either for an individual, or especially across individuals, the reliability of the data can generally be regarded as very respectable and fully adequate for use in most of the statistical analyses that are carried out with such data. Further, . . . substantial confidence can be placed in the typical statistical summarizations of such data.

Other points rating systems

Various other derivatives of points rating schemes are also in use for special purposes. These include the job information matrix system of Stone and Yoder (1970), the job element method of Primoff (1973) (cited in McCormick, 1979), and the position analysis questionnaire (McCormick *et al.*, 1972; Frank, 1982), but since they have little application to the matter at hand they are not discussed further here. A special, and comprehensive, computer-aided task inventory system known as the work performance survey system (WPSS) has also been developed by Gael (1983). Still others, such as the time-span of discretion approach (the Glacier Project), the decision-banding approach, the direct consensus method, the guide-chart profile method, and the Urwick–Orr profile method, all discussed by Livy (1975), are primarily concerned with methods of establishing fair and just pay scales, and again have little or no application here. Bartley (1981) in *Job Evaluation: Wage and Salary Administration* has also presented as extensive outline of job evaluation techniques, again oriented primarily to the establishment of equitable pay scales, while Ulery (1981) has provided idealized job descriptions for some 75 different types of personnel at the managerial, professional and foreman levels, although without any indication of the demands placed upon these incumbents in terms of time, physical or mental load, or other.

Chapter 4
Physical demands analysis

It was emphasized in the previous chapter that the ability of a worker to perform a job is dependent on the physical demands of that job and the capacity of the worker to perform the tasks that go to encompass the job. Thus, to establish whether a worker is fit to do a job it is necessary to determine the demands that the job makes upon a worker, or the abilities required of a worker to do that job. Although related, these are different processes from those described in Chapter 3 under 'Classification', 'Points rating' and other forms of job analysis. The purpose, and hence the orientation, are different. Whereas conventional and structured job analyses are intended to discriminate amongst jobs and tasks for such purposes as allocation of pay, comparison and grouping of jobs and so on, a physical demands or physical abilities analysis is intended to define those characteristics of a job which impose specific demands on workers such that, without change, they might find them beyond their capacity. Physical demands analysis is described in this chapter and physical abilities is described in Chapter 5.

Principles of physical demands anslysis (PDA)

The concept of defining job demands is relatively new, although in principle it goes back to the work of the Gilbreths in the 19th and early 20th century, who determined that jobs could be broken down into quasi-standardized component tasks, which in turn could be broken down into components and subcomponents, each requiring a particular form of activity which could be identified and timed. In so doing they laid the groundwork for what came to be known as time and motion study, with its various derivatives. Time and motion study, however, is concerned with developing preferred modes of work, standardizing these, determining the amount of time and skill required to complete a standard job, and hence the pay to be given to the worker.

Physical demands analysis, on the other hand, requires a different orientation. The techniques owe their origins to the work of Hanman (1946) who made an early attempt at matching the capacities of workers with the demands of their jobs. In particular he recognized specific physical demands such as lifting, squatting, standing and so on, as well as the environmental stresses of working in wet quarters, high atmospheric pressure and the like, and suggested matching requirements against demands so that a position commensurate with the candidate's physical capacities could be found.

The essence of his approach lay in defining some 80 characteristics of physical ability or work capacity which he could measure in objective terms such as the permissible hours of specific activity (e.g. standing or walking), the permissible weight of lifting and carrying during a working day, the type of activity permitted or not permitted (e.g. running, kneeling) and so on.

Criteria for use of PDA

A simple listing of physical demands is inadequate for use as a base on which to predicate the required functional capacities of the worker. Hogan and Bernacki (1981), basing their approach on the '*Uniform Guidelines for Employment Selection Procedures*' of the US Federal Government (US Civil Service Commission, 1978), note that the physical qualifications for screening out candidates must be (a) job-related, (b) refer only to critical or essential job components, and (c) be consistent with business necessity and safe performance of the job. The 'guidelines' in fact interpret job-relatedness to mean that employers must confine their assessments to those that predict an applicant's ability to perform critical, but not necessarily all, aspects of the job. The inability to perform unimportant or sporadic tasks within jobs increases the probability of rejecting otherwise qualified applicants who could have been employed with minimal accommodations.

Approach to assessment

A job analysis commonly begins with a review of job descriptions available in the workplace itself, or in a reference such as the *Dictionary of Occupational Titles* (US Department of Labor, 1977), or a volume such as *Job Descriptions in Manufacturing Industries* (Ulery, 1981), or other national occupational title classifications, each of which provide some familiarity with the jobs in question and will lead to a more comprehensive analysis. The next stage would commonly involve direct observation of the worker performing the operations under consideration. These operations are documented by the analyst or personnel specialist in the form of behaviour statements.

With respect to the observational process, the 'Guidelines' state:

> This method involves analyzing jobs by observing workers performing jobs and interviewing workers, supervisors, and others who have information pertinent to the job. It is the most desirable method for job analysis purposes because it (a) involves firsthand observation by the analyst; (b) enables the analyst to evaluate the interview data and to sift essential from nonessential facts in terms of the observation; and (c) permits the worker to demonstrate various functions of the job rather than describing the job orally or in writing.
>
> The analyst uses the observation-interview method in two ways; (a) He observes the worker on the job performing a complete work cycle before asking any questions. During the observation he takes notes of all the job activities, including those he does not fully understand. When he is satisfied that he has accumulated as much information as he can from observation, he talks with the worker or supervisor or both, to supplement his notes. (b) He observes and interviews simultaneously. As he watches, he talks with the worker about what is being done, and asks questions about what he is observing, as well as

> conditions under which the job is being performed. Here, too, the analyst should take notes in order that he may record all the data pertinent to the job and its environment.
> The interview process is subjective — a conversational interaction between individuals. Communication is a two-way process. Therefore, the analyst must be more than a recording device. The amount and objectivity of information he receives depends upon how much he contributes to the situation. His contribution is one of understanding and adjusting to the worker and his job.

In this connection it is emphasized that it is the responsibility of the analyst, and not the worker or supervisor, to acquire information and to structure the situation. Employees should be encouraged to describe the job in their own terms which the analyst subsequently translates or interprets in a standard form. At the same time, the worker and/or supervisor, and not the analyst, should be regarded as the authority on the job content, but the analyst should be alert for visual or other information indicating a discrepancy between the statements of the worker and/or supervisor and the apparent job situation. Such discrepancies should of course be clarified. It is helpful to back up the observational process with a video recording or with still photography.

Recording of data

Recording of the acquired data is vital. Ultimately the data will be presented in standardized terminology, commonly on standardized forms. Initial recording can be done in the form of rough notes and diagrams in free form which are subsequently made into rough draft on standard forms before final analysis. Alternatively a rough draft may be made directly on to specimen forms for subsequent refinement. The former approach is more suitable for completion of some of the more complex forms which will be discussed later, provided the analyst has a good working knowledge of the required content of these forms. The latter is useful for some of the simpler approaches. The assembled task analyses, augmented by any other information deemed significant, are then assembled in logical format in a structured task bank, which is a comprehensive list of tasks and their actions, along with the outcomes of task performance.

Criticality of tasks

Once the overall analysis is completed, it is necessary to determine which of the tasks must be considered as essential or critical. Again the views of incumbents and supervisors are of vital importance. These, however, can be augmented by measures of, for example, frequency of performance, importance to the outcome, consequence of error and so on. At this time there is little in the way of objective or even quasi-objective assessment of criticality. Hogan and Bernacki (1981) consider using rating scales for critical dimensions, having defined a cut-off score above which tasks are evaluated as being critical to the job. In one project a list of physically demanding tasks serving as a task bank was constructed from which each task was evaluated by incumbents and supervisors on seven-point scales of required physical effort, frequency of performance and importance of task performance. Means, standard deviations and standard error of the mean were calculated. To be considered critical, the mean rating of a task had to exceed a defined upper limit point, and/or had to be performed more frequently than the average task, or be rated more important than average.

Reliability and validity

Hogan and Bernacki (1981) have examined the requirements of reliability and validity of PDA data. They note that reliability can be established statistically by the use of intraclass correlations of quantitative ratings made by job incumbents and supervisors. Ratings typically used are scale scores from evaluations of tasks on critical/important dimensions. Consensus established by a panel of individuals knowledgeable about the job, such as analysts, subject matter experts, supervisors and incumbents, may also provide evidence of reliability.

The validity of a job analysis is rarely questioned and even less frequently ascertained. Rater consensus, including agreement among panel members that a task or set of tasks is required for job performance is a common though erroneous demonstration of validity. Preferred methods for demonstrating validity are (1) retrospective and (2) empirical validity. Retrospective validity involves analysis of documents certifying that the work is completed. From completed work, it is possible to proceed backwards to determine what tasks were necessary to accomplish the work. Empirical validity is a statistical process that attempts to demonstrate a relationship between test scores or ratings and an objective measure so that future scores can be used to predict, with specified assurance, that measure.

Techniques of PDA

While PDA can be of value in defining the requirements for selection and hiring of all types of employee it has to date found its greatest application in assessing requirements for the handicapped. Consequently most of the very limited literature in the field is found in relation to rehabilitation and vocational training for disabled persons. Several agencies have developed formal systems for the evaluation and recording of the working environment for this purpose. Several of these systems are examined here. One of the most comprehensive methods is 'physical demands job analysis'.

Physical demands job analysis

The physical demands job analysis system was develped by Lytel and Botterbusch (1981) for the Stout Rehabilitation Institute of the University of Wisconsin. Although it was developed specifically for evaluating job demands for the placement of handicapped persons, it is of enormous value in defining the demands of jobs for all purposes. To define their requirements Lytel and Botterbusch (1981) conducted an extensive analysis of nearly 500 jobs from which they developed a set of comprehensive questionnaire forms for outlining in detail the demands of tasks. The tasks are analysed in terms of environmental and social conditions, postures required in conducting the task requirements for handling objects and moving objects, the needs of speech and hearing, driving and machine control, visual demands, strength demands, requirements for walking and standing, the needs for communication, manipulation and mobility, along with the occurrence of special actions and the need for extended heavy demands. Each set of requirements is broken down in detail with specific questions that can be answered by check mark, but there is also opportunity for expanding on the answers and providing descriptive information. An accompanying Manual describes the requirements for each category, and defines in detail the different terms in use.

Each of the items involved is clearly defined in the Manual. It is not feasible to reproduce the Manual here, but Appendix A, taken from the work of Lytel and Botterbusch (1981) presents a completed set of forms for the simulated task of dishwasher which indicates the scope and substance of the method. It will be clear that the task of completion of the necessary forms is cumbersome, tedious and requires detailed knowledge of the Manual and considerable skill in its application. Any person wishing to use the technique should obtain and carefully peruse a copy of the Manual before adopting the method. The Manual includes seven forms on which different aspects of the analysis are recorded: the Summary Form, the Environmental and Social Conditions Form, the Job Tasks Form, the Task Analysis Form, the Visual Demands Checklist, the Strength Requirements Form, and the Physical Barriers Form.

The Environmental and Social Conditions Form provides a record in a standard and defined format of such conditions as indoor or outdoor work, temperature and humidity, noise and vibration, various defined hazards, moving objects, slippery floors and clutter, lighting, elevated surfaces, exposure to various noxious substances, and use of protective equipment. The defined social conditions refer to working relationships with supervisors and others, the occurrence of shifts, and contact with belligerent persons.

The Job Tasks Form allows for a brief written description of the major tasks in the job, sequentially presented, and including a record of the percentage of total time spent on each task, and whether the task is critical. Each of these tasks is then analysed on the Task Analysis Form which details the following items: job elements, tools and work aids, machines and equipment, most common posture, lifting heights and weights in manipulating objects, requirements for handling and finger dexterity, requirements for carrying, pushing, and pulling, speech and hearing, control placement and driving, as well as infrequent actions. This level of analysis is normally adequate for identifying potential problem areas. When problem areas are identified a more elaborate ergonomic analysis may be required.

A special form is used to evaluate visual requirements, including visual communications such as reading, visual measurements and mobility, and the effects of reduced vision on mobility in the work area.

Special consideration is also given in the Strength Requirements Form to the degree of strength required in lifting, as well as to the duration of walking, standing and sitting, and the requirement for extended or heavy physical demands on the shoulders, back, whole body or voice. Note is taken of the amount and duration of driving.

The Physical Barriers Form gives information on access to and within plant facilities such as building entrances, parking, elevators, restrooms and water fountains.

Each item on each of the forms is identified by a numbered box. The Summary Form brings all the boxes together in a classified structure and provides an overall picture of the entire analysis. Space is also allowed for comments and recommendations by the analyst.

Priest and Roessler (1983) describe the use of the method for a project with the following objectives:

1. Identify high turnover, hard-to-fill or high employment potential positions in local business and industry.
2. Identify the major tasks and production level standards of the jobs.
3. Institute vocational training in those positions in a local sheltered workshop (where required) using job analysis data to create simulated work sites.
4. Train individuals in the positions in the workshop to ensure that they can meet required job standards.
5. Recommend selected individuals for employment in participating industries.
6. Provide follow-up assistance to trainees and employers.

In the initial stages of the project five jobs with three different companies were analysed, including food processing (two positions), industrial sewing, garment inspection and computer operation. The time to analyse a job averaged 20 min with an additional 15 min required to evaluate each plant facility.

The AET approach

AET is an acronym for Arbeitswissenschaftliches Erhebungsverfahren zur Tatigkeitsanalyse and refers to an ergonomically based method of physical demands analysis originally developed by Landau (1978) under the direction of Rohmert (1985; Rohmert and Landau, 1983) at the Institut fur Arbeitswissenschaft der Technischen Hochschule, Darmstadt, Germany. Rohmert points out that since it is the aim of all ergonomic efforts to harmonize the influences resulting from the interaction of the workplace, the work content, the equipment, the working environment and the work organization with human physical and mental abilities and psychosocial needs it is necessary to examine these existing or planned work systems in the light of human stress and the resulting strain. He argues that the description of work systems that is necessary to analyse and compare these interactions requires a standardized linguistic organization which can be employed and understood both by ergonomists and experienced practitioners. AET is an attempt to achieve this result by combining engineering, physiological and behavioural–psychological elements. It was developed to evaluate work systems in which the worker is essentially involved in a production process or in rendering a service.

Table 4.1. Schema of AET content (after Rohmert, 1985).

Part A — Work system analysis

1. Work objects
   - 1.1. Material work objects (physical condition, special properties of the material, quality of surfaces, manipulation delicacy, form, size, weight, danger)
   - 1.2. Energy as a work object
   - 1.3. Information as a work object
   - 1.4. Man, animals, plants as work objects
2. Equipment
   - 2.1. Working equipment
     - 2.1.1. Equipment, tools, machinery to change properties of work objects
     - 2.1.2. Means of transport
     - 2.1.3. Controls
   - 2.2. Other equipment
     - 2.2.1. Displays, measuring instruments
     - 2.2.2. Technical aids to support human sense organs
     - 2.2.3. Work chair, table, room
3. Work environment
   - 3.1. Physical environment
     - 3.1.1. Environmental influences
     - 3.1.2. Danger of work and risk of occupational disease
   - 3.2. Organizational and social environment
     - 3.2.1. Temporal organization of work
     - 3.2.2. Position in the organization of work sequence
     - 3.2.3. Hierarchical position in the organization
     - 3.2.4. Position in the communication system
   - 3.3. Principles and methods of remuneration
     - 3.3.1. Principles of remuneration
     - 3.3.2. Methods of remuneration

Part B — Task analysis

1. Tasks relating to material work objects
2. Tasks relating to abstract work objects
3. Man-related tasks
4. Number and repetitiveness of tasks

Part C — Job demand analysis

1. Demands on perception
   - 1.1. Mode of perception
     - 1.1.1. Visual
     - 1.1.2. Auditory
     - 1.1.3. Tactile
     - 1.1.4. Olfactory
     - 1.1.5. Proprioceptive
2. Demands for decision
   - 2.1. Complexity of decision
   - 2.2. Pressure of time
   - 2.3. Required knowledge
3. Demands for response/activity
   - 3.1. Body postures
   - 3.2. Static work
   - 3.3. Heavy muscular work
   - 3.4. Light muscular work, active light work
   - 3.5. Strenuousness and frequency of movement

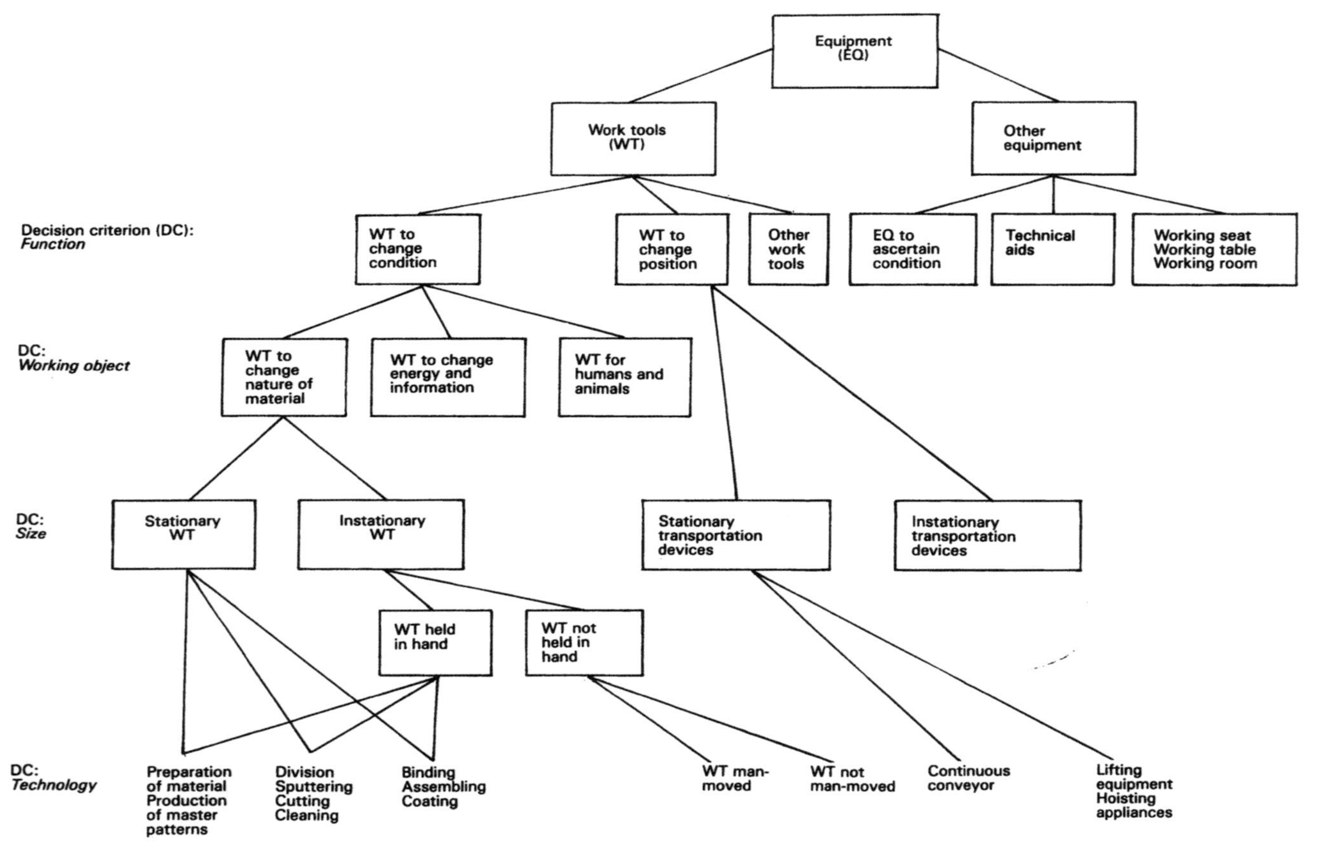

Figure 4.1. Hierarchical organization of AET (Wagner, 1985)

Table 4.2. AET item 205: stress from static work — fingers, hands, forearm (after Rohmert, 1985).

Item no.	Item	Item text
205	D	*Stress by static work* Problem: During what period of the shift is the worker exposed to stress from static work? Catchwords and explanation: The term static work implies a long-term (4 s) muscular effort which does not result in a movement of the body (contrary to dynamic work). Static work is therefore not measurable in the mechanical sense. During static work a muscular effort can take place only due to the exertion of an external force and also because of the effort required to bear the weight of the body extremities Body regions exposed to stress: finger, hand, forearm Characteristic: muscular effort without support of body weight Example: seizing and holding objects of work; operating a keyboard Code to be used: Duration code (D) 0: does not apply or applies very infrequently 1: under $\frac{1}{10}$ of working (shift) time 2: under $\frac{1}{3}$ of working (shift) time 3: between $\frac{1}{3}$ and $\frac{2}{3}$ of working (shift) time 4: more than $\frac{2}{3}$ of working (shift) time 5: nearly uninterrupted during working (shift)

Application of AET requires no measurement but is achieved, as in other systems, by observation and interview. In essence the approach resolves the total work system under consideration into individual elements while describing and scaling their interdependencies. The analysis proceeds from a model of the job and records all relevant aspects of the work object (i.e. the item on which work is taking place), the equipment, the working environment, and the task, as well as the job demands. The observation/interview quantifies all phases of the defined stress according to their duration, severity, sequence and temporal distribution within the working shift.

The analysis is based on completion of a complex set of comprehensive questionnaires containing a total of 216 items. These items are derived from a classification of the work system as illustrated in Table 4.1. Although the table as presented outlines more than merely physical demands the remainder is included for completeness.

Part A of the schema contains some 143 items out of the total of 216. These describe work objects, work tools, equipment, and the physical organizational, social and economic aspects of work. As an example, Figure 4.1 shows the hierarchical organization from which questions are developed pertaining to equipment.

Part B contains 31 items pertaining to the definition of work objects, while Part C defines human responses along the lines of a model of human activity based on concepts of perception, decision making and response/activity developed by Welford (1965).

Each of the 216 items includes a description of a potential problem using what Rohmert refers to as 'catchwords', or specific terminology, followed by an explanation of the significance of the item and specification of a code under which it might be classed. Table 4.2 illustrates the occurrence of stress from static work using the stress occasioned by static work of the fingers, hand and forearm as an example.

As noted in Table 4.2 code D refers to duration. Other codes are as follows:

Code A: defines presence or absence of stressors,
Code F: defines item frequency in the course of a shift,
Code I: defines importance/criticality of item to job,
Code S: defines severity of stress.

In practice the analyst records the coded scores/ratings for each item in a computer processable proprietary document available from the German publishers Huber Verlag, who in fact also offer a statistical evaluation of the data.

H-AET supplement

Further application of the principles used in AET have been developed in the method known as H-AET (Rohmert, 1985) which takes into account, again only from observation and interview, some of the anthropometric limitations and activity components that place demands on the worker in the workplace. The components that are considered are presented in Table 4.3.

Part A is concerned with factors which influence muscle strength, such as gender and bodily structure, as well as the location of the work object and the equipment relative to the worker. Part B1 is concerned with factors pertaining to static and dynamic work. Static work is scaled by special codes referring to the intensity, duration and direction of static muscular activities. Postural work is coded in terms of taxonomies for standing and sitting which refer to the position of the head, trunk, upper and lower limbs. Part B2 is based on a taxonomy of movements as seen from a quasi-static point of view. The motion categories of arms, hands, fingers, legs and feet, as well as the twisting movements of hands and trunk, are scaled in various ranges relative to the strain incurred, which in turn are determined by superimposition of the maximum forces applicable in the vertical and horizontal fields of movement.

The results can be presented in the form of direct analysis using coding sheets, profiles or by cluster analysis (Wagner, 1985). An example of direct analysis is shown in Figure 4.2 which lists the coded findings for the various components in a particular task.

By grouping the items into a profile it is possible to obtain a qualitative and, to some extent, a quantitative image of the activity analysed. Figure 4.3 shows a profile of the job analysed in Figure 4.2.

Table 4.3. Classification schema for H-AET (after Rohmert, 1985).

Part A — Analysis of anthropometric data for the work system

1. Analysis of worker
    1.1. Sex
    1.2. Stature (i.e. body morphology)
    1.3. Height
2. Place of assembly
    2.1. Position of place of assembly
        2.1.1. Position of assembly
        2.1.2. Activity relevance of posture
        2.1.3. Leaving place of assembly

Part B — Activity analysis

1. Static components of activity
    1.1. Static work
        1.1.1. Position of application of force
        1.1.2. Energy consumption
        1.1.3. Direction of force
    1.2. Postural work
        1.2.1. Standing
            1.2.1.1. Position of trunk/twisting
            1.2.1.2. Position of head/twisting
            1.2.1.3. Position of application of force
            1.2.1.4. Bracing areas
        1.2.2. Sitting
            1.2.2.1. Sitting facilities
            1.2.2.2. Position of trunk
            1.2.2.3. Position of head
            1.2.2.4. Position of application of force
            1.2.2.5. Bracing areas
2. Dynamic components of activity
    2.1. Partial dynamic work
        2.1.1. Categories of motion
        2.1.2. Elements of motion
        2.1.3. Length of travel
        2.1.4. Frequency of motion
        2.1.5. Position of application of force
        2.1.6. Energy consumption
        2.1.7. Direction of force

By using a multivariate cluster analysis it is possible to establish a similarity of links between various jobs and create a job dendogram. Figure 4.4 presents a dendogram of 91 jobs involving operation, control, checking and monitoring.

As with other techniques of analysis, it is desirable to be thoroughly familiar with the manual before attempting to use the process and an English version is available (Rohmert and Landau, 1983).

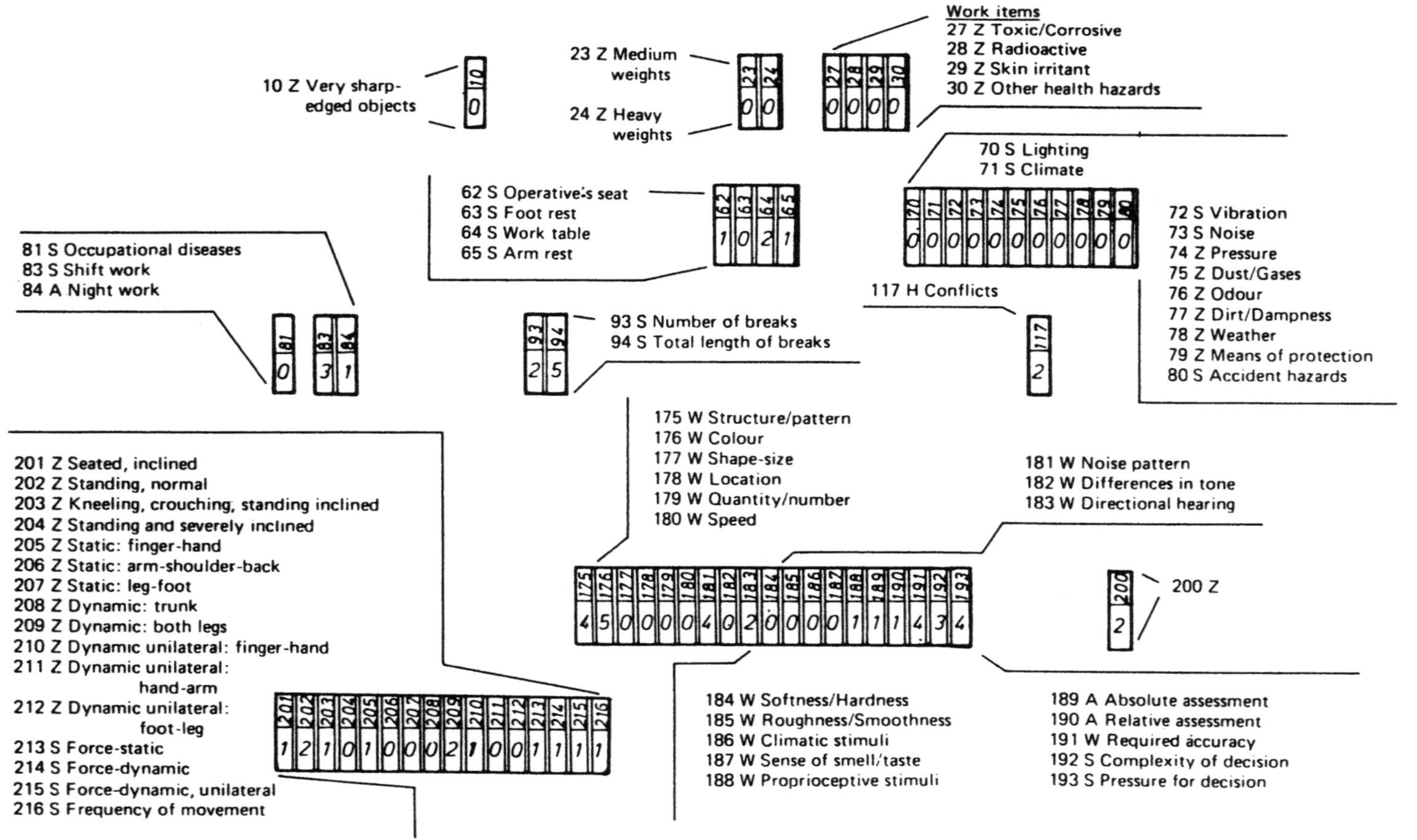

Figure 4.2. Direct analysis of components of a task using a coding sheet (Wagner, 1985)

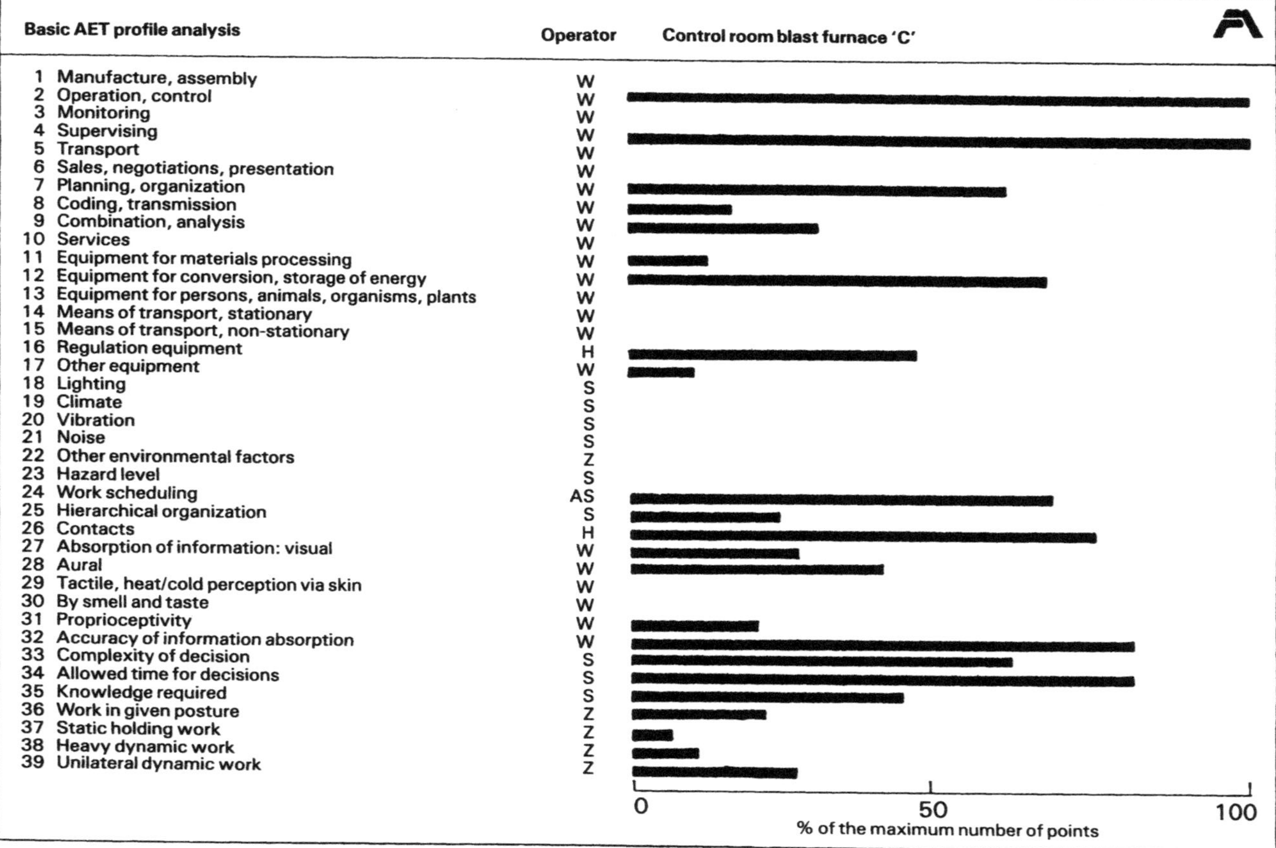

Figure 4.3. Profile of job analysed in Figure 4.2 (Wagner, 1985)

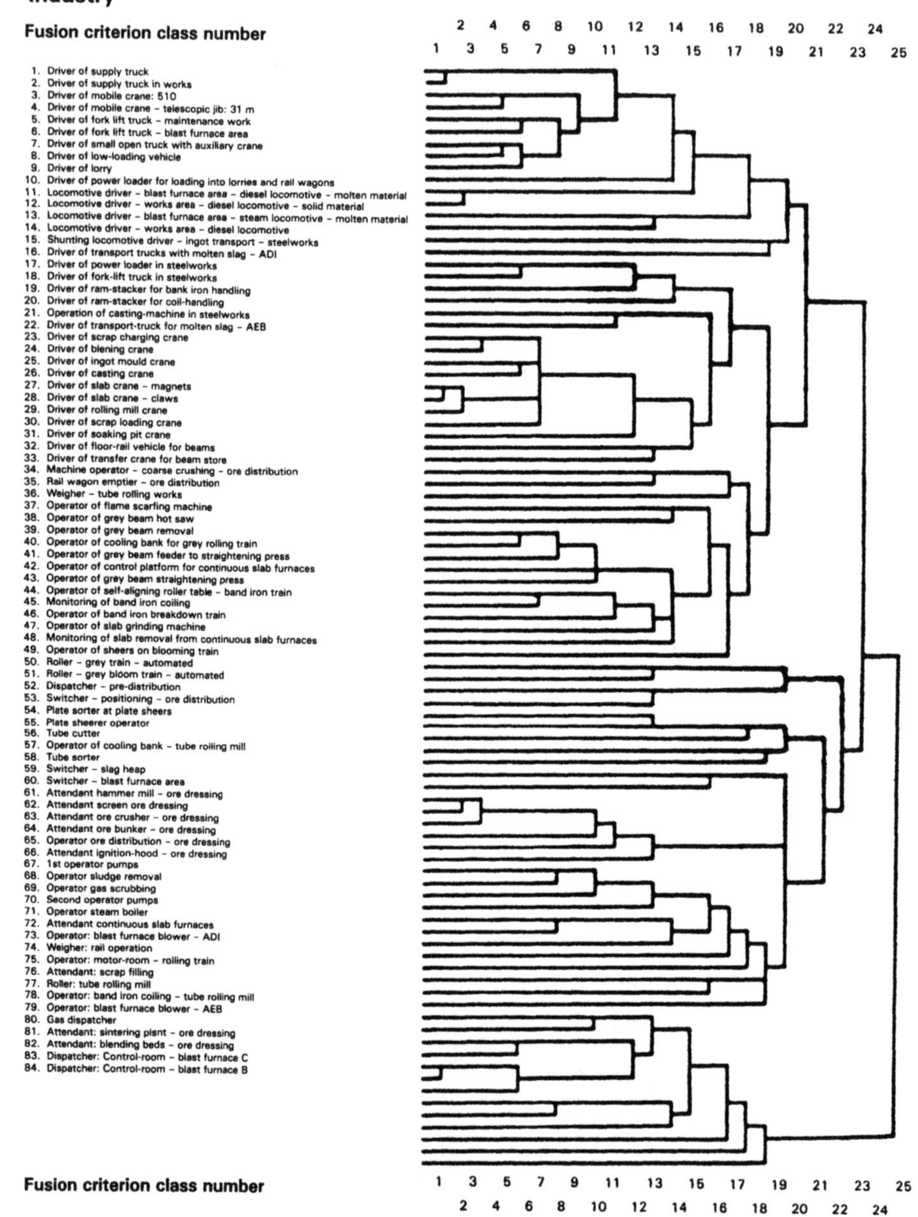

Figure 4.4 Dendogram of 91 job involving operation, control, checking and monitoring (Wagner, 1985)

Job profile method

Wagner (1985) also describes a job profiling method for assessment of work demands, developed by Renault in Europe, which is designed particularly for use with short-cycle repetitive work requiring medium accuracy. The analysis takes account of the factors shown in Table 4.4.

Each of the 27 criteria so determined is evaluated by way of a five-point scale, as shown in Table 4.5.

Coding procedures are available for each of the 27 criteria, for example factor C,

Table 4.4. Factors utilized in Job Profiling (Wagner, 1985).

Job design			Height–distance	1
			Supply–removal	2
			Space limitation, Accessibility	3
			Controls–signals	4
Safety factor		A	Safety	5
Ergonomic factors	Physical environment	B	Thermal environment	6
			Noise environment	7
			Artificial lighting	8
			Vibration	9
			Air cleanliness	10
			Appearance of workplace	11
	Physical load	C	Most common posture	12
			Most awkward posture	13
			Work effort	14
			Working posture	15
			Handling effort	16
			Handling posture	17
	Mental load	D	Mental processes	18
			Level of alertness	19
Psychological and sociological factors	Autonomy	E	Individual autonomy	20
			Group autonomy	21
	Relations	F	Relations independent of the work	22
			Relations connected with the work	23
	Repetitiveness	G	Repetitiveness of cycle	24
	Work content	H	Potential	25
			Responsibility	26
			Interest of work	27

Sitting	- Hands below heart and trunk upright	1	
	- Trunk inclined forwards (15–30°) - Trunk inclined to one side (15–30°) - Torsion of trunk (15–45°) - Hands at head level	2·5	
	- Hands at heart level, arms outstretched	3	
	- Trunk severely inclined forward (30–45°) - Trunk severely inclined to one side (30–45°)	4	
	- Torsion of trunk (45–90°) - Hands above head	4·5	
	- Trunk severely inclined backwards and hands above head (*)	5	
Standing	- Hands below heart, trunk upright	2	
	- Trunk inclined forward (0–15°)	2·5	
	- Trunk inclined forward (15–30°)	3	
	- Trunk inclined to one side (15–30°) - Torsion of body (15–30°) - Hands at head level	3·5	
	- Trunk inclined forward (30–45°) (*) - Trunk severely inclined to one side (30–45°)	4	
	- Trunk inclined forward, hands at head level - Both knees bent	4·5	
	- Trunk inclined forward, arms outstretched (*) - Trunk severely inclined forward (>45°) (*) - Trunk severely inclined backwards, hands above head - Hands above head	5	
Kneeling or squatting	- Normal kneeling posture	4·5	
	- Kneeling, hands above head etc. - Squatting	6	

Figure 4.5. Pictogram of common postures of use in AET (Wagner, 1985)

physical load. Within factor C, criterion 12, common posture, is evaluated by means of the pictogram shown as Figure 4.5. The stress values determined from Figure 4.5 are then evaluated for the proportion of the cycle during which each posture is adopted according to Table 4.6.

The same procedure is used for evaluation of criterion 13, awkward posture, using the values of Table 4.7. The work effort, criterion 14, is established by combining both force and duration (Table 4.8). Criterion 15, working posture, is determined from the aforementioned Figure 4.5, while criterion 16, handling effort, is assessed on

Table 4.5. Job profile rating scale (after Wagner, 1985).

Value	Significance
5	Very arduous, or very dangerous — for priority improvement
4	Arduous or dangerous in the long term — to be improved
3	Acceptable — to be improved if possible
2	Satisfactory
1	Very satisfactory

Table 4.6. Stress values of different postures (P1) in terms of proportion of cycle (% Tc) during which each posture is adopted (Wagner, 1985).

T1 in % Tc / P1	20 to <40	40 to <60	60 to <80	80 to 100
1	1	1	1·5	2
2	2	2	2·5	3
3	2·5	3	3·5	4
4	3·5	4	4·5	5
5	4·5	5	5+	5+

Table 4.7. Stress values of different postures (P2) in terms of proportion of cycle (% Tc) during which each posture is adopted (Wagner, 1985).

T2 in % Tc		10 to <20	20 to <40	40 to <60
P2 f/h	10 to <30	30 to <60	60 to <120	120 to <180
3	2	2·5	3	3·5
4	2·5	3	4	4·5
5	3	3·5	4·5	5

Table 4.8. Stress values for work effort (force and duration) in terms of proportion of cycle (% Tc) during which effort is expended (Wagner, 1985).

T3 in % Tc	<10	10 to <20	20 to <40	40 to <60	60 to <80	80 to 100
E1 (kg) f/h	<30	30 to <60	60 to <120	120 to <180	138 to <240	≥240
<1	1	1	1	1	1·5	2
1 to <2	1	1·5	2	2·5	3	3·5
2 to <5	1·5	2	2·5	3	3·5	4
5 to <8	2	2·5	3	3·5	4	4·5
8 to <12	2·5	3	3·5	4	4·5	5
12 to <20	3	4	4·5	5	5	5+
≥20	4	5	5	5	5+	5+

Table 4.9. Assessment of handling effort in terms of weight (p), distance (d), and hourly frequency (F) (Wagner, 1985).

F (f/h)	< 10			10 to < 30			30 to < 60			60 to < 120			120 to < 180			180 to < 240			≥240		
p(kg) \ d(mm)	< 1000	1000 to 3000	> 3000	< 1000	1000 to 3000	> 3000	< 1000	1000 to 3000	> 3000	< 1000	1000 to 3000	> 3000	< 1000	1000 to 3000	> 3000	< 1000	1000 to 3000	> 3000	< 1000	1000 to 3000	> 3000
< 1	1	1	1	1	1	1	1	1	1	1	1	1	1	1	1	1·5	2	2·5	2	2·5	3
1 to < 2	1	1	1	1	1	1	1	1	1·5	1	1·5	2	1·5	2	2·5	2	2·5	3	2·5	3	3·5
2 to < 5	1	1	1	1	1	1·5	1	1·5	2	1·5	2	2·5	2	2·5	3	2·5	3	3·5	3	3·5	4
5 to < 8	1	1·5	2	1·5	2	2·5	2	2·5	3	2·5	3	3·5	3	3·5	4	3·5	4	4·5	4	4·5	5
8 to < 12	1·5	2	2·5	2	2·5	3	2·5	3	3·5	3	3·5	4	3·5	4	4·5	4	4·5	5	4·5	5	5
12 to < 20	2	2·5	3	2·5	3	3·5	3	3·5	4	3·5	4	4·5	4	4·5	5	4·5	5	5	5	5	5
≥20	3	3·5	4	3·5	4	4·5	4	4·5	5	4·5	5	5	5	5	5	5	5	5	5	5	5

Table 4.10. Assessment of handling posture in terms of height, and distance from operator's feet as determined from Figure 4.6 (Wagner, 1985).

F in f/h (*)	<10	10 to<30	30 to<60	60 to<120	120 to<180	180 to<240	≥240
2	1·5	2·	2·5	3	3	3·5	3·5
3	2	2·5	3	3·5	4	4·5	4·5
4	2·5	3·5	4	4·5	5	5	5
5	3	4	5	5	5	5	5

(*) Stress value of postures

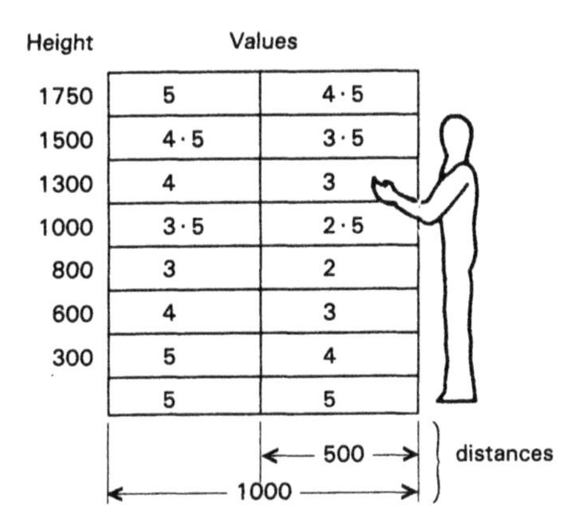

Figure 4.6. Grasp and release postures identified in terms of height and distance from position of operator's feet (Wagner, 1985)

the basis of the weighting table (Table 4.9), with particular reference to the weight handled (seven different ranges from less than 1kg to over 20 kg), distance over which the weight is moved (less than 1 m, 1–3 m, and over 3 m), and the hourly frequency. Handling postures are evaluated (Table 4.10) on the basis of height and distance from the position of the operator's feet as shown in Figure 4.6.

With respect to AET and job profiling, Wagner (1985), having used both extensively, concludes that for analysing cyclical and fairly simple activities the job profile method is advantageous, but AET, because of its flexibility and applicability to computer analysis gives better results for better tasks.

Essential PDA method

Essential PDA (Ontario Ministry of Labour, 1981) is the term given by the Ontario Ministry of Labour to a simplified but effective method of recording the physical

Ontario Ministry of Labour

Physical Demand Checklist

Date	Job Title	Contact
Branch		Telephone No. Ex.

	Physical Demands	Check if Needed	Pounds if Needed	Frequency	Essential Function Yes	Essential Function No	Possible Accommodation
STRENGTH	LIFTING include pushing & pulling effort while stationary						
	CARRYING include pushing and pulling effort while walking						
	FINGERING: right						
	left						
	HANDLING: right						
	left						
	REACHING: shoulder						
	below						
	above						
	GRIPPING: minimum						
	moderate						
	maximum						
MOBILITY	THROWING						
	SITTING		X				
	STANDING		X				
	WALKING		X				
	RUNNING		X				
	CLIMBING		X				
	STOOPING		X				
	CROUCHING		X				
	KNEELING		X				
	CRAWLING		X				
	TWISTING		X				
SENSORY/PERCEPTUAL	HEARING: conversation		X				
	other sounds		X				
	VISION: far		X				
	near		X				
	colour		X				
	depth		X				
	READING		X				
	WRITING		X				
	SPEECH		X				
ENVIRONMENT	INSIDE WORK		X				
	OUTSIDE WORK		X				
	HOT		X				
	COLD		X				
	HUMID		X				
	DRY		X				
	DUST		X				
	VAPOUR, FUMES		X				
HAZARDS	MOVING OBJECTS		X				
	HAZARDOUS MACHINES		X				
	ELECTRICAL HAZARDS		X				
	SHARP TOOLS, ETC.		X				
	RADIANT ENERGY		X				
	SLIPPERY FLOORS		X				
	CLUTTERED WORKSITE		X				
CONDITIONS OF WORK	TRAVELLING		X				
	WORKING ALONE		X				
	WORKING INDEPENDENTLY BUT IN A GROUP		X				
	INTERACTION WITH PUBLIC		X				
	EQUIPMENT/MACHINERY VEHICLE OPERATED		X				

Flexibility within Department to accommodate handicapped persons:

Accessibility to person using wheelchair:

Other comments:

1921 MOS(6/81)

Figure 4.7. Essential physical demands checklist (Ontario Ministry of Labour, 1981)

demands of various jobs. Although it was developed specifically for use in placing impaired and disabled workers it has useful applications in determining the physical demands of a job for general purposes.

In practice the preparation for the analysis, followed by observation and interview, are conducted as previously described for other methods. The results, however, are recorded on a one-page form, which, although vastly curtailed in comparison with those used in other methods, provides much of the information that might be required for suitable placement. The form is shown as Figure 4.7. Several of the columns require further explanation, as follows:

Pounds if needed: Where appropriate, the maximum number of pounds to be moved for each factor should be recorded to ensure that a worker will have the ability to meet the maximum demands.

Frequency: Despite its name this item refers to the number of hours, or fractions of an hour, that the item is needed.

Essential function: This column indicates whether a factor should be modified.

Possible accommodation: This column is completed during or after interview with a handicapped applicant. Only those factors which have been previously noted as not limiting should be considered.

While this system is useful as a general summary it suffers from obvious limitations in its scope and it is non-quantitative. It does, however, provide a useful guideline for a preliminary review of physical demands.

Graphic and notational systems

Job analysis, of course, is concerned with the evaluation of the physical demands of the job in its totality. Many of those demands, however, involve posture, movement and the application of force, each or all of which result in bodily strain. Various attempts have been made to record the demands of force and motion in either graphic symbology or some form of coded notation. A review of these processes is presented by Baleshta and Fraser (1986), and in a Master's Thesis by Baleshta (1986) from which much of the following is derived.

Photography

Photography, either cinematic or video, has been widely used in recording of activity, with subsequent analysis either frame by frame, or in slow, or ultra-slow motion, or as an overall picture. Indeed the Galbreths used photography in their time and motion studies. It is also used in methods-time measurement (MTM) work (Salvendy and Seymour, 1973) and in a wide variety of biomechanical studies, including that of Smith and his colleagues (1982) who used a super high-speed camera capable of 500 frames per second for analysis of motion in tasks where joint markers were placed on subjects who were filmed during lifting and lowering actions.

Holzman (1982) also described a photograph method for estimating effort which involved recording a task on video or film followed by frame by frame analysis at speeds dependent on the work cycle time. He suggested dividing the work cycle into 100–200 equal time intervals for this purpose. Force and posture data were coded for each frame, with effort being estimated on a Borg scale (see Chapter 5). Effort data were then entered into a computer and plotted over the cycle.

Photography, however, has limitations. It is difficult to derive meaningful quantitative data, and the results may be distorted by the camera angle and unsuitable film speeds.

Posture coding

Postures adopted during tasks have been described graphically and coded numerically by Priel (1974) who defined 14 'limbs' with 14 corresponding joints and who advocated two planes of view in order to describe limb positions in three dimensions. From the data acquired he prepared a 'posturegram' for each posture which defined the position of each joint affected, in three planes (frontal, lateral and horizontal), at any of nine angles, and any of nine levels.

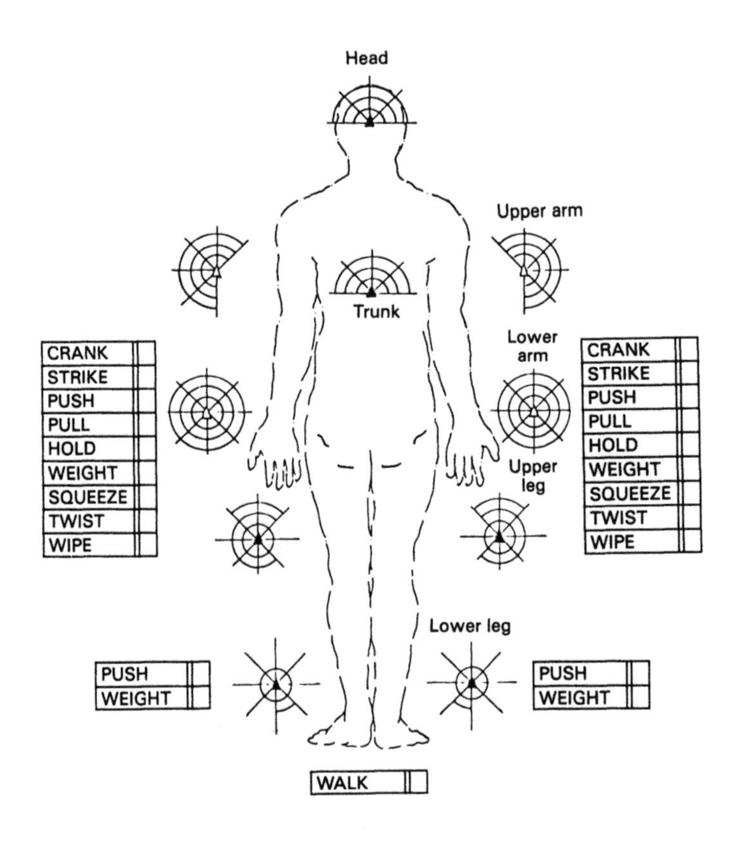

Figure 4.8. Posture targetting diagram showing standard position (*Corlett et al.*, *1979*)

Probably the most useful and simple method of recording posture was developed by Corlett and his colleagues (1979). This procedure enables a posture to be recorded from observation of a pre-arranged group of 10 'targets', which are shown in Figure 4.8 and refer to the part of the body adjacent to them.

To represent departures of the body part from the standard position in the horizontal plane a mark is made on a target according to its radial displacement, the straight-ahead position being vertically upwards on the target. Displacement in the vertical plane is recorded by treating the concentric circles as a scale of 45°, 90° and 135°, respectively from the target centre. Any forces or actions by a limb are indicated by a check mark in the adjacent panels.

Symbolic notation

Coding of motion is a more complex procedure than coding of posture. Postural coding is concerned with static states while motion is kinematic. The original systems for the coding of motion were derived from the needs of choreography. At least four of these have been in use, of which the most common is Labanotation, developed by Rudolph Laban who began work on it in 1928 (Hutchinson, 1977). The record is presented on a staff, as in music, which contains columns corresponding to different body parts. The direction of movement in lateral and sagittal dimensions is coded by the shape of symbols, while vertical directions are coded by symbol shading. Time is recorded along the vertical axis of the shaft. Different movement times are shown by the length of the symbol. Time can be related to musical beats or to any predetermined time intervals.

The Benesh system developed in 1947 by Rudolph Benesh uses a matrix to represent the human body. Ledger lines, as in music notation, are used to record extensions above and below the body. Symbols represent lateral and sagittal movements while vertical direction is indicated by the position of the symbol on the five-line stave (McGuiness-Scott, 1983).

The Eskol–Wachmann method was introduced in 1958, and is not limited to dance (Eskol and Wachmann, 1958). It is based on the concept that movements of limbs about joints describe spheres. The movements are described via limb endpoint coordinates on the surface of the sphere and the distance from the centre of the sphere to the coordinate points. These movement data are recorded in a column with vertical positions representing body parts.

The Laban concept was used by Patla and his colleagues (1984, 1985) in developing a set of symbols to be utilized in the assessment of gait, and later in the assessment of a mining task. The system is computer-mediated and is programmed in BASIC language. On a screen or printout preformed symbols describing anatomical function in a precise location are presented on a preformed chart held in the computer memory. The system, however, defines postures and does not record continuous motion.

Another development along the same lines was presented by Baleshta and Fraser (1986) for use with arm motion. Their system uses similar computer programmes to define some 53 symbols representing anatomical sites, action at joints, functional performance such as grip and wringing motion, and characteristics such as joint

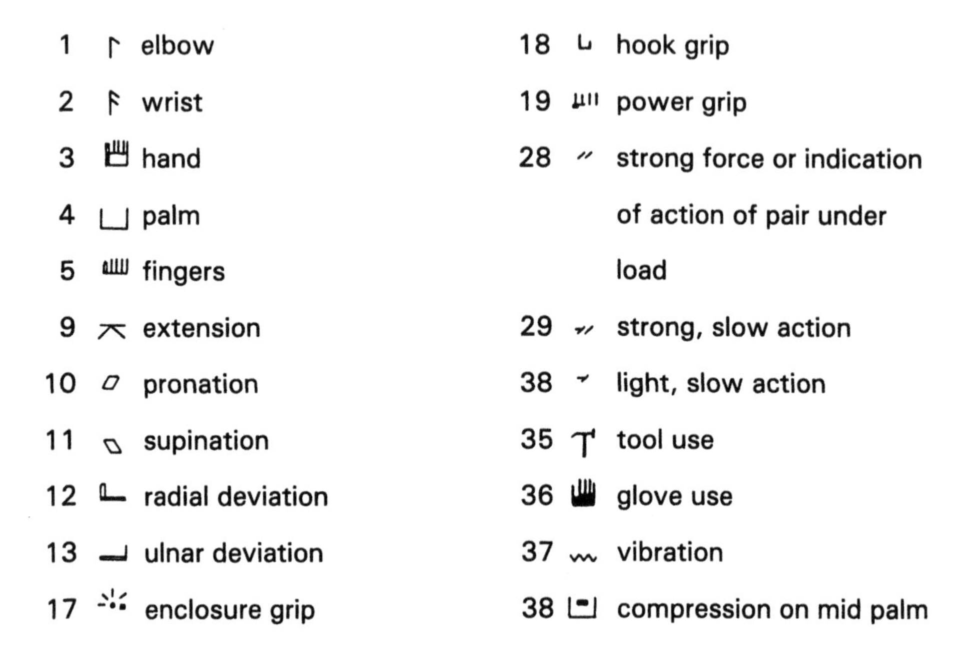

Figure 4.9. Symbols used to define arm action (Baleshta and Fraser, 1986)

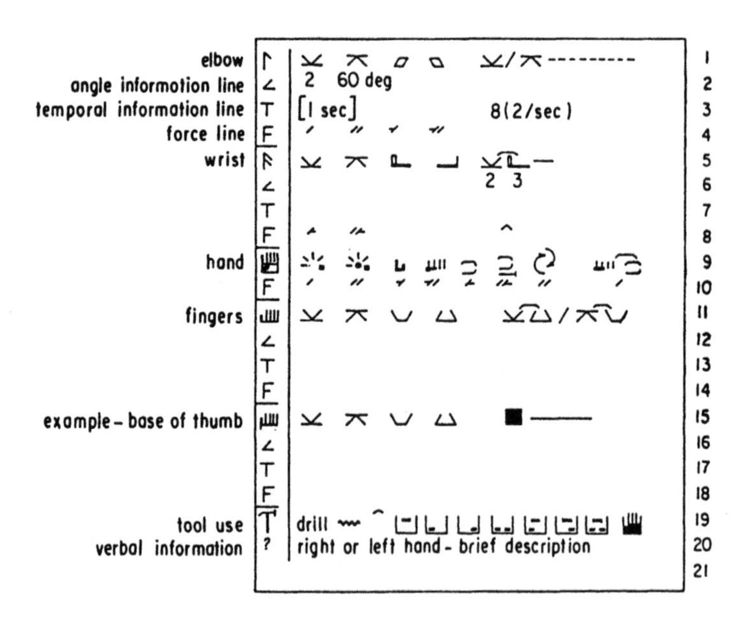

Figure 4.10. Description of arm motion using coded symbols (Baleshta and Fraser, 1986)

angle, force, duration of action, tool use and so on. Some of the symbols are illustrated in Figure 4.9. The symbols are presented on specified lines on a chart (Figure 4.10).

A notation for a given movement comprises several components. Each action is encoded by a symbol representing one or more anatomical movements, and presented on a line on the chart. As the action changes new symbols are added. Time is on the abscissa. There is also provision for linking symbols should more than one be required to encode the action of a particular joint at a particular time. Information, such as force, frequency of repetition of an action, or duration of an action, are recorded on special lines on the chart. Tool use, and other factors of significance, such as gloves, vibration, impact forces, palm compression are also notated, and there is provision for distinguishing between static and dynamic work.

Unfortunately all the symbolic notation systems described are complex in execution. It is probable that they serve no better purpose than the equally laborious task of manually recording movement by description and diagram.

Biomechanical component of task demands

As implied or noted in the discussion of the approaches to the assessment of task demands outlined above, where a lifting or carrying task is required it is necessary to determine:

— the range and maximum weight of lifts,
— the number of lifts,
— the frequency and duration of lifting,
— the type of lifting techniques required,
— the type of carry,
— the length of carry.

Various attempts have been made to encompass these and other factors into some meaningful recommendations. These have ranged from the empirical recommendation of the International Labour Office (ILO, 1962) to the complex formulary developed by the US National Institute for Occupational Health and Safety (NIOSH, 1981). The ILO recommendation is given in Table 4.11.

In the *Work Practices Guide to Manual Lifting* by NIOSH (1981) it is noted that lifting a 5 kg compact load, that is where the mass centre of gravity of the load is within 50 cm of the ankles, can create compressive forces sufficient to cause damage to older lumbar discs. As the load mass is moved horizontally away from the body a proportional increase in the compressive force on the low back is created. Thus, even

Table 4.11. Permissible weight of lift (kg) for specified age groups (ILO, 1962).

	Age (years)					
Sex	14–16	16–18	18–20	20–35	35–50	over 50
Male	15	19	23	25	20	16
Female	10	12	14	15	13	10

light loads have to be handled close to the body. Furthermore, when a load is lifted from the floor additional stresses are exerted on the low back due to the body weight moment when stooping to pick the load. The postures used to lift loads from the floor can exert complex and relatively unknown stresses on the low back when lifting. Specific instructions as to the safe posture to be used are then necessarily complex, reflecting such factors as leg strengths, load and load size. Additional problems are created when lifting loads asymmetrically, that is by one hand at the side, or with the torso twisted. This can impart complex and potentially hazardous stresses to the lumbar column, as can dynamic forces imparted by rapid or jerking motions.

In the light of these and other problems, and on the basis of numerous epidemiological, psychological, physiological and biomechanical studies of lifting, NIOSH (1981) developed a model which reflects the physical demands of the job under certain specific circumstances and can be used for determining lifting standards under these circumstances. The model is intended to apply to a situation where there is a smooth two-handed symmetrical lift directly in front of the body with no twisting during the lift. The object lifted should be no wider than 75 cm; the lifting posture should be unrestricted with good coupling between the hand and the object, and between the feet and the floor; the ambient environment should be favourable. A number of variables are included in the model as follows:

- — object weight (L) (kg),
- — horizontal location (H) of the hands at the origin of the lift, measured forward of the body centre line or midpoint between the ankles,
- — vertical location (V) of the hands at the origin of the lift measured from the floor level (cm),
- — vertical travel distance (D) from the origin to the destination of the lift (cm),
- — frequency of lift (F), as average number of lifts per minute,
- — duration of lifting period, assumed to be occasional (less than 1 h), or continuous.

For the purposes of the model, jobs are classified as *infrequent*, namely either occasional or continuous lifting less than once every 3 min; *occasional high frequency*, namely lifting one or more times per 3 min for a period of up to 1 h; or *continuous high frequency*, namely lifting one or more times per 3 min continuously for 8 h.

On the basis of the foregoing, NIOSH defined a predictive formula to establish lifting limits. Two such limits are in fact defined, one being the maximum permissible limits (MPL), and the other the action limit (AL).

The MPL, within the restrictions defined above, is based on four criteria, namely that:

1. Musculoskeletal injury incidence and severity rates increase significantly in populations when work is performed above the MPL.
2. Biomechanical compression forces on the lumbosacral discs are not tolerable in most workers at lifting levels above the MPL, that is over 650 kg.
3. Metabolic rates among most individuals working above the MPL will exceed the recommended 5 kcal min^{-1}.
4. Only 25% of men and less than 1% of women have the strength to perform above the MPL.

The AL is based on different levels of the same criteria, namely that:

— musculoskeletal incidence and severity rates increase moderately in populations exposed to lifting conditions described by the AL,
— compression force on the lumbosacral disc within the AL (i.e. up to 350 kg) can be tolerated by most young healthy workers,
— metabolic rates for workers within the AL will not exceed 3·5 kcal min^{-1},
— over 99% of men and 75% of women can lift loads described by the AL.

Three different levels are thus defined as follows:

1. Those above the MPL are unacceptable and require engineering controls for prevention of injury.
2. Those between the AL and the MPL are unacceptable without either engineering or administrative controls or both.
3. Those below the AL are considered to represent nominal risk to most industrial workers.

To meet the above requirements a formula has been defined for the AL which is then presented as a maximum load in kilograms for application in the criterion conditions described earlier. The MPL is considered to be three times the value of the AL, that is: MPL = 3 (AL). The actual formula is as follows:

$$AL = 40(15/H)\ (1 - 0{\cdot}004\,|V - 75|)\ (0{\cdot}7 + 7{\cdot}5/D)\ (1 - F/F_{max})$$

where AL = action limit; H = horizontal location (cm) forward of midpoint between ankles at the origin of lift; V = vertical location (cm) at origin of lift; D = vertical travel distance (cm) between origin and destination of lift; F = average frequency of lift (lifts min^{-1}); F_{max} = maximum frequency which can be sustained as shown in Table 4.12.

In the foregoing formula the varying factors act as modifiers to the basic 40 kg. Thus if the H factor, namely $15/H$, is 15 cm no adjustment is necessary. Similarly, when V = 75 cm then the V factor equals unity and no adjustment is required. The D factor ranges from 1 to 0·74 as D varies from 0 to the maximum of 200 cm. A vertical travel distance of 25 cm is considered to be the minimal value allowed, and D is never set less than 25.

With respect to the F factor, when, for example, the lift originates below 75 cm and is performed continuously throughout the 8-h day then, according to Table 4.1 2, F_{max} would be 12. Thus if the frequency of lift (F) were 6 lifts min^{-1} the F factor would be (1 – (6/12)) or 0·5, thereby halving the amount of load lifted.

The demands of a job however, can also be expressed in terms of the motions of the

Table 4.12. Maximum sustainable lifting frequency for use with NIOSH lifting guidelines (NIOSH, 1981).

Duration (h)	$V > 75$ cm Standing	$V <= 75$ cm Stooped
1	18	15
8	15	12

body segments involved in its execution, that is, in terms of three fundamental characteristics — displacement, velocity and acceleration. The determination of these, unfortunately, is a relatively complex matter and should be done by specialists, using, for example, the photographic or electromechanical techniques described by Ayoub (1972).

These forces and torques can be utilized to calculate the job demand power at different articulation joints and to provide an accurate definition of the job demands as applied to the musculoskeletal system.

A more widely used and simpler approach is to determine the average work output in terms of changing the body posture and position of the tool or object handled. The average work output is given by the expression Fd/T, where F is the force due to body weight and the weight of the work tool or object, d is the distances through which the force has acted, and T is the time taken to move the mass through distance d. The limitations, of course, are obvious: an average value may or may not be representative of peak demand in terms of either work output or strength.

Another approach to the estimation of work output is modelling. If the characteristics of the job, such as the workplace geometry, the initial and final points of motion, the weight and size of the object to be handled and so on are known, along with the basic anthropometric data of the worker an optimization model can be developed. From the model two classes of job motions can be developed: (i) the motion associated with the best performance, that is, with the job performed at minimum cost in mechanical energy, and (ii) the motion associated with worst performance, that is, with the job performed at maximum cost in mechanical energy. Two limits for work output and peak strength can then be computed, from which the expected range of physical and physiological response can be predicted.

Legal Criteria

Bearing the foregoing in mind, Thompson and Thompson (1982) have examined the criteria and characteristics of job analysis that can be acceptable by the US courts. Firstly, they note that for court acceptance a formal job analysis is mandatory. Furthermore, the analysis must be done on the job for which selection is required. It must be reduced to written form. The data required should be collected from several current data sources such as interviews with incumbents, supervisors and administrators, training manuals and other pertinent publications, questionnaires and checklists. Data collected from individuals should be from a large enough sample to be representative of every position the test is intended to cover.

Tasks, duties and activities must be identified and included in the job analysis. All of the tasks must be covered, and the relative degree of competency necessary for entry level must be specified. The courts also stress the identification of tasks as a prerequisite to an acceptable job analysis.

This chapter has examined the physical demands of the job determined by observing and measuring where feasible the nature of the workplace and the requirements thereby placed upon the worker. The physical abilities and the functional capacity of the worker to meet such demands are considered in the next two chapters.

Chapter 5
Physical abilities analysis

In the previous chapter discussion has centred on the approach to evaluation of human physical capacity in the light of work demands, with the ultimate objective of ensuring a match between the employee and the job. Thus, by evaluation of the medical fitness or physical capacity of the applicant to meet the critical demands of the job, in terms of physiological, biomechanical or psychological attributes, a profile of the applicant can be defined to meet the profile of the job.

Instead of emphasizing the physical demands of the job, another approach lies in defining the physical abilities required of the worker. These in turn can then be used as a guideline or yardstick against which the capacities of an employee candidate can be measured.

This approach, known as physical abilities analysis (PAA), is dependent on the quantification or pseudo-quantification of human perception of effort, and is derived largely from the work of Fleishman and his colleagues.

Before going on to outline the techniques, however, it is desirable to discuss some of the underlying concepts.

In a paper concerned with defining a taxonomy of human performance, Fleishman (1975) observes that task definitions vary greatly with respect to their breadth of coverage. At one end of this dimension are definitions that view the task as the totality of the situation imposed, including ambient stimuli. The other end of the dimension is represented by definitions that treat a task as a specific performance such as depressing a button when a light appears.

With respect to the type of analysis considered in Chapter 5, Fleishman recognizes what he terms the 'behaviour description approach', in which categories of tasks are formulated on observation and description of what operators actually do while performing a task, for example soldering, using tools and so on, as represented by the USES and the Functional Job Analysis type of assessment.

He compares this with what he terms the 'behaviour requirements approach' which emphasizes the cataloguing of behaviours that are assumed to be required in order to achieve certain criterion levels. The human operator in this approach is assumed to possess a large repertoire of processes that will serve to intervene between stimulus events and responses, as utilized in the position analysis questionnaire (PAQ) type of evaluation. Typical of the functions used to discriminate behaviour and differentiate among tasks are scanning functions, short-term memory, long-term memory, decision making and problem solving.

As noted, however, PAA is different from the foregoing. It is dependent on an attempt to quantify the perception of effort or exertion experienced while doing a task.

Quantification of perception of effort

In an extensive paper Gamberale (1985) observes that perceived exertion can only be measured indirectly by the use of self-reporting techniques in which the subjective response to exertion is recorded in quantitative or semi-quantitative terms by the person experiencing the effort. The applicability of subjective criteria in assessment, however, depends on (i) the type of subjective reaction observed, (ii) the way in which the reaction is observed and recorded, (iii) the extent to which the reaction varies systematically in different operations, (iv) how well the reaction correlates with work intensity and performance, and (v) how well it correlates with physiological and neurological events. No single subjective reaction, measurement method or experimental strategy is more adequate than others in every condition and for all purposes. He defines four techniques that have been used for such evaluation, namely, ratio scaling, category scaling, rating scaling and acceptability scaling. Each is examined below.

Ratio scaling

The earliest attempt to quantify the functional relationship between perceptual magnitude and physical dimensions was undertaken by Stephens (1957) who showed that it was in the form of a power function, namely:

$$S = KI^n$$

where S = magnitude of sensation, I = intensity of physical stimulus, K,n are parameters characteristic of the experience and can be determined empirically.

The expression can also be presented in logarithmic terms to the base 10, and graphed as a straight line, thus:

$$\log_{10}S = n\log_{10}I + \log_{10}K$$

An exponent of I greater than 1 indicates that the intensity is a positively accelerated function of the physical stimulus; an exponent less than 1 is indicative of a negative acceleration.

The concept can be used in the estimation of intensity. In ratio estimation, which is the most common, a subject estimates the percentage magnitude of a stimulus as compared with a known standard. In magnitude estimation the subject matches a number directly to the perceived magnitude in the light of a standard which has previously been assigned a numerical magnitude.

Various studies have been done to determine the intensity of perceived effort. Borg and Dahlstrom (1960) using a bicycle ergometer found an exponent of 1·6, while Stevens and Mack (1959) found an exponent of 1·7 with subjects squeezing a handle.

Category scaling

The concept of category scaling is also based on a presumed relationship between perceived intensity and effort. No attempt is made, however, to compare with a known standard or estimate a ratio. In this situation each end of the scale is defined, for example, 1 might be considered the level at which pain is first perceived, and 5 the level at which it becomes intolerable, with other levels interspersed in between. Various authors quoted by Gamberale (1985) obtained a linear relationship between time on an isometric task and subjective estimates of pain or perceived intensity, although it has been suggested (Kinsman and Weiser, 1976) that in fact the subjects in these cases were estimating time rather than effort.

Rating scales

Rating scales are the most commonly used devices for scaling effort. Although superficially similar to category scales they are different in that in the latter there is no assumption made regarding the distance between different values, while with rating scales the values are considered to be ordinal.

The most commonly used rating scale is that developed by Borg (1962), and known as the 'rating of perceived exertion' (RPE) scale. The scale was developed on the basis of empirical data from work on the bicycle ergometer where a linear relationship was shown between heart rate and RPE, and consequently between work load and RPE, since heart rate varies directly with workload.

The scale comprises some 15 values, some of which are verbally defined, the others being interpolated, thus:

6
7 very, very light
8
9 very light
10
11 fairly light
12
13 somewhat hard
14
15 hard
16
17 very hard
18
19 very, very hard
20

Many studies have been conducted using the RPE scale. One of the more significant (Pandolf *et al.*, 1972) showed that heat load did not have an impact on the RPE comparable to the physiological strain it produces, and that the presence of heat load will create conditions where a person will overestimate their physical endurance.

Acceptability scaling

The fourth technique is that of acceptability scaling. In this case the subject is invited to adjust a workload, for example, the weight of a lift under controlled circumstances, to a level which he/she considers to be the maximum tolerable without straining or becoming unusually tired. In the standard procedure the subject on one occasion will start with a heavy weight and on another a light weight; the two results will be averaged. The technique has been widely used by Snook (1978) and by Ayoub and his colleagues (1979), in determining lifting capacity, although as yet there is no general agreement on its validity.

There appears to be consensus that the best method of recording a relationship between perceived exertion and its stimulus is the ratio-scaling method, and specifically the magnitude estimation technique. In practical settings, however, where absolute comparisons between different workstations are needed, or when interest is focused on inter-individual or intra-individual comparisons, the use of a rating scale can have clear advantage compared with magnitude estimation.

Gamberale (1985) concludes that perceived exertion should be considered as constituting a summing up of the influence from all structures under stress during exercise. At the same time he recognizes that there is no single objective counterpart for this perceptual phenomenon.

Physical factors involved in perception of effort

Fleishman and his colleagues have adopted a similar view of perceived exertion as a summation of stress phenomena in reports published over the last 25 years or more. In particular, in a discussion outlining the background to the development of a physical abilities assessment procedure, Fleishman and Hogan (1984) note that basic to a large class of human task performance is physical proficiency, but conclude from various studies (Fleishmen, 1962; Fleishman *et al.*, 1961a,b) that neither a general physical proficiency factor nor a general strength factor exists in performing physical work. Fleishman, *et al.*, (1984) also notes that work physiologists regard effort in terms of energy expenditure, expressed as a respiratory, metabolic or cardiovascular variable; ergonomists and industrial engineers consider effort in terms of work output variables including motions used, time elapsed, fatigue factors, weights, distances and amount of materials handled; and psychologists use psychophysical responses to different workloads and scaling or related variables such as task variability. Consequently it should be feasible to integrate all three approaches on the assumption that the performer's subjective report of perception associated with physical work can be used to obtain information about physiological responses to work and workload, for ultimate use in determination of functional capacity and work demand.

Various studies were undertaken to validate this concept. Fleishman (1964) investigated and compared human performance on more than 100 well-known and diverse physical proficiency tests. The results of these supported the hypothesis that perceived effort could be correlated with physical proficiency. Statistical correlations indicated that there were nine factors or abilities which could account for the variance in evaluating physical proficiency. Of these, four were associated with strength, and two with physical flexibility. The others comprised co-ordination, equilibrium and stamina, respectively. Each is described in the following paragraphs.

Dynamic strength: Dynamic strength is defined as the ability to exert muscular force repeatedly or continuously over time. This type of strength is related to muscular endurance and the resistance of muscles to fatigue. It involves the power to propel, or to support or move the body repeatedly or for prolonged periods. It is not, however, found in single static movements, nor in sudden explosive movement, nor does it imply a resistance to aerobic fatigue. Tasks requiring this ability include pull-ups, push-ups, rope climbing and other tasks where the body is moved or supported, usually with the arms.

Trunk strength: Trunk strength is a derivative of dynamic strength and is characterized by the resistance of trunk muscles to fatigue over repeated use. It describes the degree to which the abdominal and lower back muscles can support part of the body repeatedly or continuously, as in obtaining or recovering from postures involving flexion at the hip, such as leg-lifts or sit-ups.

Static strength: Static strength is the force that a person exerts in lifting, pushing, pulling or carrying external objects. It represents the maximum force a person can exert for a brief time period where the force requires some major effort. Because it involves a brief time period resistance to fatigue is not a feature as it is in dynamic strength.

Explosive strength: Explosive strength is characterized by the ability to expend a maximum amount of energy in activities involving one or a series of maximum thrusts. It is distinguished from other types of strength by the necessity to mobilize energy, largely anaerobic, in a burst of effort rather than continuously. It is used in tasks requiring rapid, forceful propulsion with the feet and hands as in jumping or throwing.

Extent flexibility: Extent flexibility involves the ability to extend the trunk, arms and/or legs through a range of motion in the frontal, sagittal or transverse planes. The emphasis is on the range of movement encompassed. It is involved in tasks that require suppleness, such as in reaching and stretching.

Dynamic flexibility: Dynamic flexibility is contrasted with extent flexibility in that it involves the capacity to make rapid, repeated, flexing movements in which the capacity of the muscles to recover is critical, for example, where a person is continuously and rapidly rotating joints or body parts through a range of angle, as in, say, operating a screwdriver.

Gross body co-ordination: Gross body co-ordination is the ability of the body to maintain balance in either an unstable position or where forces are acting against the stability for the body either standing or moving. This can be a factor in working from ladders, stools or scaffolds.

Stamina: Stamina can be considered as representative of cardiovascular endurance. It enables the performance of prolonged bouts of aerobic work without experiencing fatigue or exhaustion. It is best indicated by oxygen consumption. The factor is associated with total body involvement rather than localized muscle function, or with long-term aerobic work rather than short-term anaerobic work, and is responsible for sustained performance.

Application to job analysis

The foregoing concepts outline some of the physical factors that go to determine the experience of effort. Fleishman *et al.* (1984) have taken these concepts and used them in the analysis of the abilities required in work. To develop the procedure Hogan and Fleishman (1979) identified for this purpose 99 tasks for which the metabolic costs had been calculated. These tasks were assembled into lists of occupational and recreational tasks. Each of the tasks in these groups was then rated by personnel specialists who were not aware of the metabolic costs. The rating was done on a modified Borg scale in which the standard 15 values had been reduced to 7 while preserving the same grading. A relationship with a confidence level of $P = 0{\cdot}01$, highly significant, was found between the perceived level of effort and the measured metabolic cost. On replication of the project with raters who were not personnel specialists similar findings were obtained.

From this experiment a revised effort scale was developed with Borg's adjectival comments, such as 'very, very, hard' replaced by behavioural task anchors such as 'operate a calculator', 'perform light welding', 'operate a jackhammer' and so on. Using this scale the best predictors of effort were found to be ratings of stamina, explosive strength, static strength and dynamic strength. The work was extended by Hogan *et al.* (1980) and Gebhardt *et al.* (1981) using a wider range of occupational tasks and two additional physiological measures of metabolic costs. A final study, also by Gebhrdt *et al.* (1981) was then conducted validating the effort scale against actual task performance in the laboratory using 20 subjects doing 24 manual materials handling tasks. The results showed a substantial agreement among raters with a correlation coefficient of $+0{\cdot}83$.

From all these various studies Fleishman *et al.* (1984) conclude firstly that the task ratings of perceived physical effort used by them are highly related to metabolic costs, ergonomic costs and ratings of physical abilities in task performance; secondly, that ratings of perceived effort are reliable for both generic and narrowly specified tasks; thirdly, that no difference in ratings is attributable to age or sex of the rater or occupational experience; fourthly, that the dimensions of the physical domain most highly related to individual perceptions of effort are the ratings of various strength and stamina factors. They also consider that a major implication of the results is that the perceived physical effort scale is a reliable and valid assessment device for predicting costs, and that it can be used reliably by both job incumbents and supervisors, with different fitness levels and experience, to rank and group diverse tasks and jobs according to common levels of physical demands.

Technique of physical abilities analysis

On the basis of the work of Fleishman and his colleagues, Nylander and Carmean (1984) developed a technique of quasi-quantitative evaluation of the abilities required of a worker to perform a given task, which in turn provided a profile of each task based on the same selection of characteristics. The technique is described in a questionnaire known as the Physical Abilities Analysis Manual (PAAM), along with a similar technique for the analysis of working conditions (Nylander and Carmean, 1984).

The approach is predicated on the assumption that human performance capacities, or abilities, dictate the probability of successful completion of a task, and thus form the basis for a system that classifies task requirements in terms of human abilities. Tasks and tools may change over time but the dimensions of human performance capacities remain relatively stable. The task requirements are described using a stable set of human capacities which underlie all performance in a wide variety of tasks but vary only in degree. Thus, performance observed on one task can be related to that observed on others and can lead to more reliable predictions about task performance for tasks which are nearly identical.

In practice, the PAA approach to task analysis requires workers or other knowledgeable raters to make judgments regarding individual differences in the degree of an ability required to perform various tasks. It is critical for a rater, whether an incumbent or a knowledgeable observer, to establish an appropriate judgment strategy if he/she is to provide reliable information about the job, and if employers are to make the correct decisions based on inferences drawn from the data. Thus, before actually assigning a rating a rater must firstly gain an understanding about the physical ability that he/she is attempting to rate — what it means and how it differs from other abilities. Once this is obtained the rater must then think of a task performed on the job that requires that ability. Grasping that relationships and assigning a numerical rating to it in accordance with guidelines is the essence of the rater's task. This concept will become clearer after the rating systems have been described.

PAAM/WCAM rating system

The rating system developed by Nylander and Carmean (1984) has been reviewed by many private corporations, as well as by state and local governments in the USA, and by federal agencies in the USA, Canada and Australia, and has been widely used in its original or modified form. Fundamentally it comprises two basic sets of questionnaires known as the Physical Abilities Analysis Manual (PAAM) and the Working Conditions Analysis Manual (WCAM). These are bound together as a set and preceded by a Biographical Data questionnaire to be completed by or on behalf of the job incumbent. The joint Manual includes extensive instructions as to how the questionnaires should be administered. It is common practice for one instructor or supervisor to administer the questionnaires to a group of workers whose jobs are being rated. The Manual provides specific instructions as to how this should be done, including standardized wording that the instructor will use in explaining the process. For example, at the beginning of the activity the instructor identifies himself/herself and his/her department and states:

> Today you will be rating the working conditions and physical ability requirements of your job.
>
> We will be using this information to develop pre-employment medical standards. These standards must relate directly to the physical demands and requirements of the job and they should promote safe and efficient job performance.
>
> The information you provide today will help us to develop job-related medical standards that will be used only for new employees in your class.
>
> Today's session will last approximately two to three hours. Before we begin, are there any questions?

The instructor goes on to describe the Biographical Data questionnaire in detail, explaining the significance of each question and indicating how the rater might mark it, and completing each question, one at a time, until the form is complete. He/she then goes on to the Working Conditions questionnaire painstakingly in the same manner until it too is complete. The same approach is made to the Physical Abilities questionnaire. Written statements are provided in the Manual for explanation of each item to be considered. The content of each of the subsystems is amplified further below. Sample pages from the Manual covering different aspects of the Biographical Data, Working Conditions Analysis and Physical Abilities Analysis questionnaires are presented in Appendix B.

Biographical Data questionnaire

The Biographical Data questionnaire provides information on such matters as sex, ethnicity, education, time in present position and number of persons supervised (if any). It also asks the appraiser to state whether he/she is suffering from impairment of sight, hearing, speech, impairment of physical ability because of amputation, loss of function or loss of co-ordination, or any health impairment which requires special education or related services.

WCAM Rating questionnaire

The Working Conditions Manual defines 34 different sets of working conditions, some of these being physical, some chemical and some social. Although the defined conditions do not completely describe the working environment, and may indeed be open to question in some instances, they are presented here, along with their definitions as given in the Manual. As will be described later, each item is graded according to terms defined specifically for each, on a scale of 0–3, but the scales are not shown here.

1. *Inside*
Working under a roof and with all sides protected from the weather (exclude motor vehicles from consideration).
2. *Outside*
Working outside exposed to the weather — heat, cold, humidity, dryness, wetness and dust (due to climate rather than other sources).

3. *Low temperature*
Working in a relatively low average degree of temperature.
4. *High temperature*
Working in a relatively high average degree of temperature.
5. *Sudden temperature changes*
Working where temperature changes of more than 10° may take place.
6. *Low humidity*
Working under conditions in which the atmosphere contains a low degree of moisture relative to temperature and air movement.
7. *High humidity*
Working under conditions in which the atmosphere contains a high degree of moisture relative to temperature and air movements.
8. *Wetness*
Contact with water at site of work.
9. *Slippery surfaces*
Working where there is a possibility of falling or losing one's footing because of slippery surfaces.
10. *Body injuries*
Possibility of cuts, bruises, sprains, fractures or amputation.
11. *High elevations*
Working above floor or ground level.
12. *Confined spaces and/or cramped body positions*
Positions in which the worker is narrowly hemmed in, or work which requires an awkward or strained position.
13. *Moving objects*
Work on or above moving machinery or equipment, in the vicinity of vehicles in motion or near any object that changes place or position whereby the well-being of the worker may be jeopardized.
14. *Vibration*
Exposure of the body, particularly the arms and legs, to sudden jerks and jars, or vibration.
15. *Noise*
Working condition in which sound is produced as part of the work process or is part of the job.
16. *Burns*
Possibility of injuries to the body caused by heat, fire, chemicals or electricity.
17. *Non-ionizing radiation*
Possiblity of exposure to radiation caused by welding flash, microwaves or sunburn.
18. *Dust*
Working in an area where the air contains quantities of fine, dry particles of earth or matter, other than free silica or asbestos.
19. *Silica dust*
Working in an area which contains free silica or asbestos dust.
20. *Allergenic*
Working in situations with the possibility of exposure to common allergy-causing

agents such as bee or wasp stings and poison oak, ivy and sumac.

21. *Toxic conditions*

Exposure to toxins; dust (other than silica and asbestos), fumes, liquids, gases (aldehydes, other than gases resulting from plastics fires; or carbon monoxide, the effects of which may be multiplied by smoking or proximity to open flame) which cause general or localized disabling conditions.

22. *Chemical irritant*

Working in situations where chemical irritants such as fires with plastics may be involved.

23. *Oily*

Using oil or grease in normal performance of work.

24. *Odours*

Working condition in which the worker necessarily comes in contact with noxious air.

25. *Explosives*

Working with or near material which, under certain conditions, is apt to rapidly burst or break up into pieces accompanied by a noise.

26. *Electric hazards*

Possibility of contact with uninsulated or unshielded electrical equipment.

27. *Ionizing radiation*

Possibility of exposure to radiation from such sources as radioactive isotopes, X-rays and other nuclear substances.

28. *Infections*

Any infections caused by micro-organisms. Rated in terms of probability of occurrence rather than actual occurrence, and severity.

29. *Air pressure*

Working under a high- or low-pressure condition caused by atmosphere or compressed air forces.

30. *Working with others*

Association with others in the course of job performance.

31. *Responsibility for persons*

Having responsibility for the welfare and lives of others.

32. *Job complexity*

The degree and depth of such factors as significant professional training, specialized knowledge and skills, variability of tasks and analytical requirements.

33. *Role ambiguity*

Lack of clarity about what others expect of you on the job.

34. *Irregular or extended work hours*

Working under conditions that cause fluctuating work hours.

Each of these conditions, as previously noted, can be rated on a scale of 0–3. Although the terminology of the rating levels is specific to each condition being rated, essentially the levels can be interpreted as follows: 0, no effect; 1, little effect; 2, moderate effect; 3, great effect.

Thus, for example, with respect to noise the scale is defined as follows:

NOISE — Working conditions in which sound is produced as part of the work process or is part of the job.
0: No effect. Is not a condition of the study job.
1: Little. Low sounds or occasional fairly loud sounds.
3. Moderate. Steady and fairly loud noises.
4. Great. Intermittent or continued loud and insistent noise.

On the other hand with respect to role ambiguity the scale occurs as follows:
ROLE AMBIGUITY — Lack of clarity about what others expect of you on the job.
0: No effect. Is not a condition of the study job.
2. Little. Rarely is it not clear what others expect of you on the job.
3. Moderate. Occasionally what is expected is not clear.
4. Great. What is expected is usually not clear.

PAAM Rating

As noted, the PAAM rating system is based on the application of quantitative scales to the type of factors defined by Fleishman and his colleagues (1984) in their study of effort. In the Manual, 22 such factors are defined, organized in groups according to the pre-established classification of strength, stamina, flexibility, body movement, dexterity (use of arms and hands) and sensation (specifically vision and hearing). Each factor in turn is then defined, described and discriminated from related factors. It is not feasible to reproduce here the content of the entire Manual. Those who wish to use the system in its entirety (and it is indeed a useful tool) should consult the original work of Nylander and Carmean (1984). The most feasible approach here is to outline the nature of the technique with suitable examples so that a reader may seek further information if desired, or may at least be familiar with the concept for future understanding.

For example, strength is considered in terms of the four factors, static strength, explosive strength, dynamic strength and trunk strength. The Manual identifies item one, namely *static strength*, as '... the ability to use muscle force to lift, push, pull, or carry objects. It is the maximum force that one can exert for a brief period of time. This ability can involve the hand, arm, back, shoulder, or leg ... '. Static strength is differentiated from other manifestations of strength, namely dynamic strength, trunk strength, explosive strength and stamina. The nature of the differences is outlined in Apendix B (Part III, item 1).

As implied in the above differentiations, other factors in the strength classification are explosive strength, dynamic strength and trunk strength. Each is defined and categorized accordingly. Two factors are considered in the stamina classification, namely stamina itself and effort. Flexibility includes extent flexibility and dynamic flexibility. The classification of body movement is applied to the factors of mobility, speed of limb movement, gross body co-ordination, and gross body equilibrium. Dexterity includes arm–hand steadiness, manual dexterity and finger dexterity.

Once the rater has understood the significance of the terms he is asked to rate his job, as he sees it, on a scale of 1–7, with 1 at the low end of the scale and 7 at the high end. Each factor has its own specific scale with descriptive anchors at each end and one or more further intermediate descriptions, as illustrated in Appendix B, with respect, for example, to static strength and manual dexterity.

With respect to sensory factor, only vision and hearing are rated. Vision includes near vision, far vision and the requirements for discrimination of colour. Hearing includes the requirement for hearing in a quiet environment, in a noisy environment, for the location of sound and for discrimination of non-speech sounds. While vision, like the other factors, is rated on a seven-point scale with descriptive anchors and intermediate guidelines which indicate the demands of vision in terms of what a worker may be able to do, such as '... read the fine print of legal journals ... ', the rating of the hearing items is based solely on criticality, with 1 representing '... not essential or critical to performance of the job ... ' and 7 representing '... essential or critical to the performance of the job ... '.

Type and number of raters to be used

Nylander and Carmean (1984) address the problem of number and type of raters, noting, that there has been some conflict in job analysis in general as to who should provide the information. Although the incumbent is involved directly with the work performed, concerns have been raised that he/she may not be able to articulate or express in writing the demands of the job. There is also a concern of 'false positives', that is inflation of the importance, frequency or difficulty of the work performed by the incumbent. They note, however, that there are a number of advantages in obtaining job information from incumbents. Firstly, they are more familiar with all aspects of the job, including infrequently performed component tasks. Secondly, estimates of task criticality defined by time spent and amount of required abilities reflect valid ratings when compared to actual samples. Finally, since the number of incumbents typically exceeds the number either of supervisors or of job analysts, the chances for reliability are increased simply from the statistical power generated from an increased sample size. They conclude that the selection of rater is an empirical issue which may be specific to both methodologies and jobs.

With respect to the number of raters required for statistical reliability in the results, they also note that the size of the sample depends on such factors as the sample design, the size of the group, and administrative consideration such as cost and the need to develop defensible medical standards for a specific classification. On this basis it is recommended that a rating group comprise a minimum of 30 persons. This does not mean that method cannot be used on individuals to obtain specific individual information; any information so obtained, however, would not be reliable for predictive purposes nor to define the needs of a group.

The sex of raters might also be considered significant. There is not sufficient evidence to draw definitive conclusions at this time, but in one study reported by Hogan and Fleishman (1979) in which 26 male and 26 female untrained examiners rated 30 occupational and 41 recreational tasks it was found that the mean task ratings for women ranged from 1·8 to 6·6 and for men from 1·6 to 6·6, suggesting that both men and women were responding similarly in their ratings.

It can be argued that the PAAM/WCAM technique does not define or describe the actual physical demands of a job in any objective terms; it merely describes the perceived effort required of a statistically determined average of a group of workers doing that job. Also, it can only be used where the job is already being done by the incumbents, and consequently cannot be used to predict what demands might be made on a worker where a new job is being designed or an old job modified; nor, of course, does it provide any way of comparing the demands of the job with any legal or otherwise specified requirements. It does, however, provide a useful and operationally oriented guide for the determination of pre-placement medical standards, and particularly medical standards that are job-oriented, legally defensible and allow for the placement or rejection of workers who might otherwise have been placed or rejected on empirical grounds for want of knowledge by the examiner of the demands of the workplace.

Thus, in use, the steps required for a selection procedure using PAAM/WCAM are as follows:

1. The employer collects information on the physical abilities required for the respective jobs from all relevant employees and prepares a PAA.
2. The health professional conducts an assessment of the candidate.
3. If diseases, abnormalities or impairments are revealed, the health professional consults the standards derived from the PAA previously conducted.
4. If the conditions required of the job do not exceed the capacity of the applicant, the applicant is acceptable.
5. If the conditions required of the job do exceed the capacity of the applicant, the PAA will assist in defining any accommodation that might be required or provide substantive evidence for unsuitability of the candidate.

Chapter 6
Functional capacity assessment

Functional capacity assessment (FCA) describes the evaluation of an employee or employee applicant in terms of their capacity to perform the essential duties of a position. FCA does not define a technique so much as a range of procedures used to evaluate fitness for work. It provides a basis for appropriately matching an individual to a specific job or general type of work from previously defined job or work demands, and may or may not include the specification of aids and accommodations to improve the match. It is the intent in this chapter to describe firstly the theory and concepts of FCA, and secondly the practical approach to its application. Its use of more than one assessment procedure distinguishes it from a standard employment medical examination (Ramanujam 1988). In addition, it encompasses access to the job site for detailed observation of the job, and also for a job trial in those circumstances where it is desirable.

For an FCA to be effective and accurate it must be done in conjunction with the physical demands analysis (PDA) of the job. Without the PDA it has no specific context and reverts to being no more than a general medical examination of the applicant, producing privileged information which belongs better with the family doctor or specialist. Without the context of the PDA, the medical assessment can only report in general terms on those medical or physical problems that show up in the examination, and may provide little information of value in determining the potential match between applicant and job.

The more detailed the PDA and FCA, the fairer is the process of selection both to employer and candidate. It is fairer to the employer because it provides a bench-mark against which to measure applicants, and it is fairer to applicants since it is confined to assessment of fitness for only those duties which are necessary for the safe and efficient performance of the job, and does not address irrelevant tasks that may bar persons with disabilities.

Even when a job duty is identified that a person cannot perform, the PDA and FCA, by virtue of their detail and positive orientation towards finding a match between what someone can do and what needs to be done, make it easier to determine the kind and extent of reasonable accommodation required in the circumstances.

Medical fitness

FCA can be considered as having two components, evaluation of medical fitness and evaluation of work capacity. To these components may be added the requirements for a simulated or actual job trial when circumstances demand. The assessment of medical fitness has been examined extensively in Chapter 2, where it was shown that with respect to fitness for work much of the traditional medical examination may be inadequate and indeed to some extent irrelevant, while its unthinking practice and irresponsible usage may conflict with contemporary concepts of human rights legislation, particularly for the employment of the handicapped. A medical (physical) examination of a candidate for employment is nevertheless still a necessary part of an FCA. The examination should, however, be a directed examination, that is, an examination which is targeted to the capacity of the worker in the light of the demands of the job. In a definitive paper, Goldman (1986) outlines the requirements of pre-placement examinations in a manner which has particular applicability to those of an FCA. She emphasizes that the examination is not a substitute for primary prevention, or the elimination of hazards, achieved by selection of a 'superman' worker who can withstand undue stress or hazardous conditions, but that the primary goal of the examination is to reveal any medical condition that might put the worker at risk or cause risk to others, so that an appropriate placement or accommodation can be made. The requirements of the examination as she defines them are ideally suited for the medical examination of the FCA and are described below.

Occupational health history

As already noted, a health history is the cornerstone of the fitness for work examination. The required history, however, is more than the general medical history which is part of a traditional examination, although the latter is also necessary. The occupational health history comprises firstly a job profile, secondly a listing of symptoms, illnesses or injuries related to past jobs, thirdly questions relevant to the requirements or potential hazards of the new job, and fourthly a listing of community and home exposures relevant to the work at hand. The requirements of the job profile, which outlines the work an applicant may have previously done, are shown in Table 6.1.

The required history of symptoms, illnesses and injuries related to past jobs is useful to avoid placing the potential employee in conditions that might precipitate recurrence. The types of question required to elicit the information are presented in Table 6.2.

Table 6.1. Job profile (Goldman, 1986).

A Workplace name, location, products manufactured
B Job title and description of operation
1. Chemical (generic) or physical form of agents
2. Operating and clean-up practices
3. Protective equipment and clothing
4. Ventilation and/or other engineering controls
5. Eating and/or smoking at the job site
C Exposure monitoring information
D Inclusion of part-time jobs and military service

Table 6.2. Illnesses and injuries related to past jobs (Goldman, 1986).

A	Do you have any health problems, symptoms, or injuries associated with your most recent or past jobs?
B	Have you ever worked with any substance that caused you to break out in a rash?
C	Have you ever worked at a task that made you short of breath, cough or wheeze?
D	Have you ever had to change jobs or work assignments because of a health problem or injury?
E	Have certain types of work caused you significant strain in your limbs (e.g. tendonitis) or back?

The general occupational history should be supplemented by inquiring into symptoms or illnesses pertinent to the new job requirements or potential hazards. For potential exposure to lead, for example, symptoms or conditions similar to those found in lead poisoning, such as anaemia, difficulty concentrating or remembering, weakness in the hands, problems in the reproductive system, might be elicited.

Potentially hazardous exposures in the home environment can be contributory to problems in the work environment. Table 6.3 indicates the type of questions that could reveal potential problems in the home and community.

Table 6.3. Home and community exposures (after Goldman, 1986).

A	Do you live very close to a factory?
B	Do you know of any community environmental health problems, e.g. contaminated drinking water or chemical spills?
C	Do you have any hobbies that involve potentially hazardous exposures?
D	Do you use pesticides in the home or garden?
E	Do you have any conditions at home that you think might affect your health, e.g. use of aerosol sprays, chemicals, cleaning agents, recent reconstructions, painting?

General medical history

While the occupational medical history is of immediate concern in placement, it provides an inadequate basis on which to select a potential employee. It is also necessary to obtain from the applicant a relevant past medical history, with particular reference to the presence or past occurrence of any medical illnesses or injuries that might place the worker at increased risk to himself or others. As Goldman (1986) points out, it might be desirable to place a worker with chronic persistent hepatitis in an area without solvent exposure or to recommend the use of a respirator with even minor exposures to solvents. A worker with epilepsy might not be placed in a job that would involve the running of large or moving machinery, although in all cases, as will be discussed later, it is necessary to take account of any relevant legal requirements and guidelines which address discriminatory employment practices by which a medical examination might select out certain protected classes.

Lifestyle history

Lifestyle is also an increasingly important factor that must be considered in any physical examination for employment. There is no doubt that the use of drugs, smoking and alcohol may place the worker and his/her fellows at increased risk from various job exposures, such as are found with smoking among asbestos workers, or machinery or vehicle handling among alcohol and other drug users. Consequently it is necessary to make careful, if perhaps discreet, enquiry into various lifestyle habits in the course of an FCA.

Physical examination

Essential to the assessment, of course, is the actual physical examination, which must be conducted by a physician knowledgeable about the conditions of work and the specific demands of the position under consideration. It is unfortunately all too easy for a physician to administer the same standard physical examination to all potential employees regardless of the needs of the job, and to complete that examination with the same set of standard biochemical or other laboratory tests, some of which may be inadequate and some irrelevant. The examination for an FCA, however, must be a directed examination, that is, one that is oriented to the particular demands of the work, and not merely a casual laying on of hands.

Hogan and Bernacki (1981) emphasize that the primary function of the examiner in this situation is one of risk assessment; that is, the physical examination is designed to determine a person's past and present state of health such that the physician may predict the likelihood of immediate or future health impairment secondary to the performance of a particular task or tasks. The outcome should be capable of validation through medical or epidemiological studies supporting that assessment. Such prediction is based on the following assumptions: firstly, that everyone is subject to risk of impairment or death; secondly, the risk for each group can be described by some average or mean risk, and thirdly, the mean risk for a group can be adjusted quantitatively to define the probable risk for an individual by estimating the effect of personal and environmental characteristics. Thus, for example, smokers chronically exposed to greater than permissible levels of asbestos can be shown epidemiologically to be at greater risk of developing cancer than smokers not exposed to that agent; consequently a smoker should not be placed in that environment. Similarly a person with early cirrhosis should not be placed in an environment where he/she could be exposed to greater than permissible levels of chloroform or similar solvents, nor, as another example, should patients with arrythmia, who might be completely acceptable in another environment, be placed in situations of potential exposure to industrial solvents. In no case, of course, should restricted placement be used as an excuse for not maintaining a safe environment. In general, if the risk potential of an applicant within a particular environment may result in impairment, and if a reasonable accommodation cannot be made to reduce the excess risk, then the individual should be precluded from working in that environment. If, however, the risk either in the individual or the environment can be reduced to acceptable levels the person can and should be allowed to accept employment in that environment.

The FCA then, begins with an occupational medical history and a general medical history, in the light of the work demands and the environmental demands. It is continued with a review of systems to determine, at least from an overview, whether there is any manifest disease or disorder which could be aggravated by the intended work or affect the safety of the applicant or others. It has been argued in Chapter 2 that there is perhaps little to be gained by the apparently healthy applicant from the traditional laying on of hands, and that equally significant information can be obtained for this purpose by the use of an appropriate questionnaire administered by qualified health personnel and reviewed by a physician where necessary. Old habits die hard, however, and it may be difficult to convince an attending physician, or even a tradition-bound management, that the physical examination of the past is no longer a cost-effective procedure. If, however, it is done, it should be limited to such procedures as are necessary to confirm or deny any real or potential impairment that might prejudice the selection of the applicant. Routine laboratory or special testing that does not directly serve that purpose is neither necessary nor desirable. Where a history, either occupational or general medical, or a systems review, reveals a potential health problem that could interfere with safety, then of course any special testing that can clarify that problem should be done. FCA, however, is not the place for development of generalized baseline information. If this is desirable, and indeed it may be, it should be sought on another occasion when an employee has been hired. Consequently, routine special testing that does not directly contribute to the decision-making process involved in hiring should not be undertaken at that time. Even on another occasion when it might be deemed desirable to acquire baseline information for management purposes, care should be taken to ensure compliance with the appropriate legislation or guidelines.

Work capacity

A physical examination can determine within broad limits the state of medical health. To meet the requirements of job matching, however, it is necessary to examine the nature and measurement of work capacity and to ensure that the capacity of the worker is compatible with the demands of the job.

While the capacity of a person to function depends on the adequacy of performance of his/her total physiological being, the capacity to do work, and particularly physical work, depends for the purposes of this analysis on the functional performance of the body respiratory, cardiovascular (including cardiopulmonary) and musculoskeletal systems. In general, the work capacity of a person is the integrative sum of the functions of these physiological systems, and can be defined in terms of the:

(a) ability to breathe,
(b) ability to transfer oxygen effectively in the lungs,
(c) ability to increase cardiac output to meet a given work load,
(d) ability to transport oxygen to the working muscles,
(e) ability to exert adequate muscular force (Ayoub, 1983).

The physiological function is expressed in terms of the capacity to utilize energy efficiently and effectively in:

(a) achievement and maintenance of posture,
(b) exertion of appropriate strength,
(c) reaching, grasping, holding, manipulating,
(d) pushing, pulling, lifting, carrying.

Comprehensive assessment of work capacity, however, is a complex procedure requiring skilled professional and technical assistance, as well as sophisticated equipment. This type of full assessment would only be necessary under exceptional circumstances where, for example, there is strong pressure to determine whether a doubtfully acceptable candidate has indeed the physical capacity to perform demanding work. An outline of some of the techniques utilised is given in Appendix C.

Physical fitness testing

For most purposes a lesser degree of sophistication is required. It is indeed no more desirable to undertake a comprehensive work capacity evaluation as part of a routine examination than it is to subject an applicant to a string of unnecessary and irrelevant laboratory tests. There may be occasions where it is desirable to assess cardiopulmonary physiology and measure static and dynamic strength. There may, on the other hand, be many situations where actual measurement of work capacity is unnecessary and all that is required is clinical assessment of whether an applicant can perform the necessary functions of his job. Indeed, it is often sufficient to obtain a broad overview of the general level of physical fitness of the applicant as a baseline on which to establish his/her specific fitness to do the job at hand, bearing in mind that the primary objective of the FCA is to establish whether the candidate is capable of doing the essential duties of the job under consideration. Where fitness testing is deemed desirable, although full work capacity assessment is unnecessary, a less sophisticated approach may suffice, as indicated later, although even this approach may be too general for many purposes. It is presented here for completeness. It must be recognized, however, that a generalized physical fitness test procedure may not be recognized in law as a valid test of work requirements unless it can be shown that the skills and abilities tested are those required by the job under consideration (see Chapter 9).

Much work has been done in the field of physical fitness testing, most of it under controlled laboratory or field conditions, often with trained or selected subjects. For practical purposes, one of the simplest and most useful sets of procedures has been developed by the Canadian Ministry of Fitness and Amateur Sport, and published as the *Canadian Standardized Test of Fitness* (Ministry of Fitness and Amateur Sport, 1986), from which the following material has been derived. Although it is recommended that the testing be conducted by a registered fitness appraiser, and preferably one who is certified in cardiopulmonary resuscitation, the essence of the procedure is presented below to allow some understanding of what is required. It is emphasized that only an outline is presented here. Many of the tests, however, can be used in isolation for the particular purpose of determining specific work capacity, although that is not the purpose of the test programme, nor is it recommended by the originators for such usage. It should be recognized that there are various warnings and caveats that should be observed before administering tests to all types of persons. An evaluator who wishes to conduct the programme according to proper procedures should be familiar with the Operations Manual and observe the requirements therein.

The testing should be conducted under reasonably standard conditions in a well-ventilated room measuring at least 5 × 6 m and removed from extraneous noise. The temperature ideally should not exceed 28°C. If the floor is carpeted a board should be placed under the weigh scale near a wall and the participant should stand on the board when the height is measured.

The participant is advised not to eat for at least 2 h before undergoing the appraisal. He/she should also refrain from caffeine beverages and smoking for 2 h, and alcohol for 6 h.

Equipment

The equipment is relatively simple, most of it being part of the normal armamentarium of a health professional's office or clinic. The remainder can be purchased relatively cheaply, or constructed according to plans provided in the Manual. It comprises the following:

— Table
— Chair with arm support for blood pressure measurements
— Stethoscope
— Sphygmomanometer
— Timer or stopwatch
— Metric tape for height measurement
— Set square
— Spring or beam scale, calibrated with known weights
— Fat calipers
— Anthropometric tape
— Ergometer steps (constructed according to plan provided)
— Cassette tape with exercise instructions and music
— Cassette tape recorder (calibrated against a metronome)
— Hand dynamometer
— Flexometer (modified Wells and Dillon, constructed according to plan provided).

Test procedures

The procedures used in testing are also relatively simple, and most of them are already familiar to occupational health professionals. The procedures comprise the following:

Anthropometry:
— Standing height
— Body weight

— Girths: chest, waist (abdomen), hip (gluteal), right thigh
— Skinfolds: triceps, biceps, subscapular, iliac crest, medial calf
Aerobic fitness:
— Steptesting
— Post-exercise heart rate
— Post-exercise blood pressure
Muscular strength, flexibility and muscular endurance:
— Grip strength, right plus left hand
— Push-ups
— Trunk forward flexion
— Sit-ups.

Each will be amplified below, emphasizing those procedures which may be less familiar to the occupational health professional.

Anthropometry

Height and weight
Weight is measured with the subject standing against a wall, on the wooden board if necessary. The set square is placed on the head and the wall is marked with masking tape at the point of contact. The weight scale is also placed on the wooden board if necessary.

Girths
Chest girth is measured at the mid-level of the sternum after normal expiration and with the tape horizontal. Waist girth is taken at the point of noticeable narrowing or at the estimated level of the 12th rib. The right thigh is measured in the standing position, feet slightly apart, with the tape 1 cm below the gluteal line.

Skinfolds
Skinfolds are measured with Lange or Harpenden fat calipers. They are measured in the erect position with the musculature relaxed to the extent feasible. The triceps fold is measured on the back of the right arm at a point midway between the tip of the acromion and the tip of the olecranon. The biceps fold is measured on the right extended upper arm over the biceps at the same level as the mid-arm point for the triceps. For the subscapular measurement, the subject stands with the shoulders relaxed and the arms by the sides. The fold is measured on a diagonal line coming from the vertebral border of the scapula to a point 1 cm beneath the inferior angle. The medial calf fold is measured with the relaxed right foot flat on a step so that the knee is at 90°. The measure is taken on the inside of the right calf just above the level of maximum calf girth.

Once the site is determined a fold of skin plus the underlying fat is grasped between the thumb and forefinger with the back of the hand facing the appraiser. Keeping the jaws of the calipers always at right angles to the body surface the contact faces of the calipers are placed 1 cm below the point where the skinfold is raised. While maintaining pressure of the fingers on the skinfold the trigger of the calipers is fully released and the measurement is taken.

Aerobic fitness testing

Assuming a resting heart rate and blood pressure have been obtained, the aerobic fitness testing procedure comprises a series of stepping sequences performed on double 20·3 cm steps to a six-count musical rhythm set by the cassette tape, with progressive increases in tempo, followed by post-exercise determination of heart rate and blood pressure.

The test is structured such that a participant begins by performing a 3 min warm-up exercise at a cadence intensity of 65–70% of the average aerobic power expected of a person 10 years older. Instructions and time signals are given by the cassette tape as to when to start and stop exercising and for the counting of the 10 s measurement of post-exercise heart rate. If a predetermined ceiling post-exercise heart rate is not attained or exceeded, the participant performs a further 3 min of exercise at 65–70% of the average aerobic power expected of his/her age. Again, if the participant does not attain or exceed the ceiling heart rate a further 3 min of stepping is performed at an intensity equivalent to 65–70% of the average aerobic power for a person 10 years younger. The participant may complete a maximum of three stepping sessions. Ceiling post-exercise heart rates are shown in Table 6.4.

Table 6.4. Ceiling post-exercise heart rates (Ministry of Fitness and Amateur Sport, 1986).

	Heart rates 10 s after	
Age	1st session	2nd session
60–69	24	—
50–59	25	23
40–49	26	24
30–39	28	25
20–29	29	26
15–19	30	27

After the participant completes the last session of stepping, determined by the post-exercise heart rate response, the post-exercise systolic and diastolic blood pressure readings are determined, initially between 0·5 and 1·0 min post-exercise, and again between 2·5 and 3·0 min.

Strength, flexibility and endurance

Strength, flexibility and endurance are sampled by measures of grip strength, push-ups, trunk forward flexion and sit-ups. Grip strength is measured by use of a hand dynamometer. The latter is a hand-held device with two handles which can be squeezed together to exert maximum force. Two trials are allowed per hand. The score is recorded as the combination of maximum values for each hand.

Push-ups are conducted according to standard procedure. For the male they are conducted with the subject lying on the gym mat and pushing up on the arms with the toes as a pivot. For the female the knees are used as the pivot point. The upper body is kept in a straight line and push-ups are performed consecutively but without a time limit.

Trunk forward flexion is measured by way of a modified Wells and Dillon Flexometer which can be purchased or constructed from plans available in the Manual. It comprises a wooden frame from which a scaled rod projects forward. A sliding marker is on the rod. The flexometer is placed on the floor and the subject sits on the floor with his/her feet on the floor and the knees straight. The subject bends forward and pushes the sliding marker to maximum reach along the scaled rod. The maximum reading from two trials is recorded.

Sit-ups are conducted with the subject lying on the mat with knees bent, hands at the side of the head and elbows forward. The ankles of the subject are held by the examiner while the subject performs as many sit-ups as possible within 1 min. During each sit-up the elbows touch the knees.

Interpretation of fitness testing

Body weight, adiposity and fat distribution

Both the amount and distribution of body fat have been identified as being closely related to morbidity and mortality data. One method of evaluating these factors lies in considering four indices, as follows:

(a) *Body mass index* (BMI): The ratio of the body weight to the square of the height.
(b) *Sum of skinfolds* (SOS): This requires the addition of the values determined for the triceps, biceps, subscapular, iliac crest and medial calf skinfolds.
(c) *Waist to hip ratio* (WHR): The ratio of the waist girth to the hip (gluteal) girth.
(d) *Sum of trunk skinfolds* (SOTS): The trunk skinfolds concerned are those of the subscapular and the iliac crest.

Percentile values for males and females of 15–69 years are given in the Manual. Some representative values for the 75th percentile subjects of 30–39 years are given in Table 6.5.

Should a person be found to have a high BMI it must be determined whether this is a result of excessive body fat content or elevated muscle mass. If the SOS value is also high there is evidence of too much body fat, with a consequent health risk. Should the BMI be unduly low, again the SOS will assist in determining if the subject has too

Table 6.5. Representative values of BMI, SOS, WHR and SOTS (Ministry of Fitness and Amateur Sport, 1986).

	BMI[a]		SOS		WHR		SOTS	
Gender	M	F	M	F	M	F	M	F
	23	21	41	55	0·84	0·72	22	20

[a]BMI, Body mass index; SOS, sum of skinfolds; WHR, waist to hip ratio; SOTS, sum of trunk skinfolds.

Table 6.6. Energy requirements in litres of oxygen per minute for different states of fitness test (Ministry of Fitness and Amateur Sport, 1986).

Stage	Males	Females
1	1·1391	0·9390
2	1·3466	1·0484
3	1·6250	1·3213
4	1·8255	1·4925
5	2·0066	1·6267
6	2·3453	1·7867
7	2·7657	

little fat. An excessive amount of fat in the trunk region has been shown to be associated with increased morbidity, for example, glucose intolerance, hyperinsulinism, blood lipid disorder and so on. Some investigators have suggested that the WHR provides a valid representation of the pattern of fat distribution. Since the SOTS provides a very direct measure of subcutaneous fat in the trunk region this value is also considered. Thus, it is possible that a subject may have acceptable BMI and SOS values, but still have excessive trunk fatness as determined by high WHR and SOTS values.

Maximum Aerobic Power

Maximum aerobic power can be predicted from the step test by using the following regression equation:

$$VO_2 \max (\text{ml kg}^{-1} \text{min}^{-1}) = 42{\cdot}5 + (16{\cdot}6 \times VO_2) - (0{\cdot}12 \times H) - (0{\cdot}24 \times A)$$

where VO_2 = average oxygen cost of last completed exercise stage (l min^{-1}) (see Table 6.6); W = body weight (kg); H = heart rate after final stage of stepping (beats min^{-1}); A = subject's age (years).

Norms and percentile scores for the predicted maximal volume of oxygen consumption are presented in the Manual. Values for the 20–29 year age group for males and females are shown in Table 6.7.

Table 6.7. Representative norms for predicted maximal oxygen consumption (l min^{-1}) by 20–29-year-old males and females (Ministry of Fitness and Amateur Sport, 1986).

Gender	Male	Female
Excellent	>57	>40
Above average	53–56	37–39
Average	43–51	35–37
Below average	40–42	32–34
Poor	<40	<31

Special testing

Vision

While many activities can be conducted by persons with varying degrees of blindness, and should be defined as such, most work requires functional vision. There is a strong correlation between visual acuity and work efficiency (Schober, 1983). Visual standards have been defined for certain specific occupations such as flying and vehicle driving, varying with the type of flying or driving required. For instance, the US Department of Transportation requires for commercial interstate freight drivers, a visual acuity of 20/40 (Snellen) in each eye and with both eyes, with a horizontal field of 70° in each eye, along with the ability to discriminate the colours of traffic signals. The American Medical Association Ad Hoc Committee on Medical Aspects of Automotive Safety recommended in 1969 (reviewed in 1972) that there be three classes of drivers, namely: Class 1, professional drivers of passenger emergency vehicles (e.g. ambulances), visual acuity 20/25 (Snellen) in each eye with correction of less than 10 dioptres; Class 2, commercial taxi drivers and truck operators, central visual acuity of 20/40 in the better eye and 20/60 in the worse eye, with a special report if correction is 10 dioptres or more; Class 3, operators of personal vehicles, 20/40 or better in one eye. There is, however, no consensus on standards for industry or general working activity, although companies may have special standards for special work, for example for colour discrimination, for precise work or for use of special optical systems. If the requirement for good vision is a significant feature of the work demand, a simple visual screening test is desirable, amplified where necessary for special purposes or for assessment of impairments, by referral to an optometrist or ophthalmologist as required.

Various visual screening devices exist, such as the Orthorater (Bausch and Lomb, Inc.), the Sightscreener (American Optical Company), and the Vision Tester (Titmus Optical Company). These devices can be used to screen for visual acuity, far and near, each eye separately, colour vision, phorias and depth perception, although the test available for depth perception may not be considered sufficiently accurate to be used as a basis for work restriction (Bond, 1975). As in other evaluation situations it may be necessary to conduct a simulated or actual work test before a full judgment can be made.

Hearing

Many work activities today are conducted in a noisy environment which may approach or even reach the level of hazard. The existence of a hearing impairment, however, should not be considered a bar to employment unless that impairment is such as to interfere with the capacity of the worker to do the job. There is still considerable discussion amongst legislative authorities, management and worker institutions as to whether a worker with a hearing impairment should voluntarily be permitted to accept or continue work in a noisy environment which could intensify his/her impairment. It is unlikely that such an argument can be resolved on a scientific basis; it is argued by some that a worker has the right to damage his/her

Table 6.8. Allowable background noise levels for hearing conservation audiometric rooms (ANSI, 1960).

Octave band Centre frequency (Hz)	500	1000	2000	4000	8000
Level (dB)	40	40	47	52	62

hearing if he so wishes. At the same time, no such controversy should ever be permitted to act as a reason for not ensuring a safe noise-free environment.

Hearing testing is best conducted in a commercial sound booth with external noise attenuation such as that recommended by the American National Standards Institute (ANSI, 1960) or equivalent, as illustrated in Table 6.8.

Audiometry may be undertaken by a manually operated device or in semi-automatic fashion. With the manual device and the subject wearing carefully fitted binaural earphones under controlled conditions, a sound is introduced into each ear separately and sequentially at each of the frequencies given in Table 6.8. The intensity is increased until the subject indicates perception; the intensity is further increased before being reduced again until the subject can no longer hear it. The average level of perception and loss of perception, in decibels (dB), is considered to be the hearing threshold at that frequency. The semi-automatic system conducts a similar test without the continued participation of an operator. All testing should be conducted by a properly trained audiometric technician or the equivalent. The results are presented in the form of an audiogram (Figure 6.1) from which the extent, and to some extent the type, of impairment can be determined. Referral to an audiologist is desirable where the average loss is greater than 25 dB over the frequencies of 500,

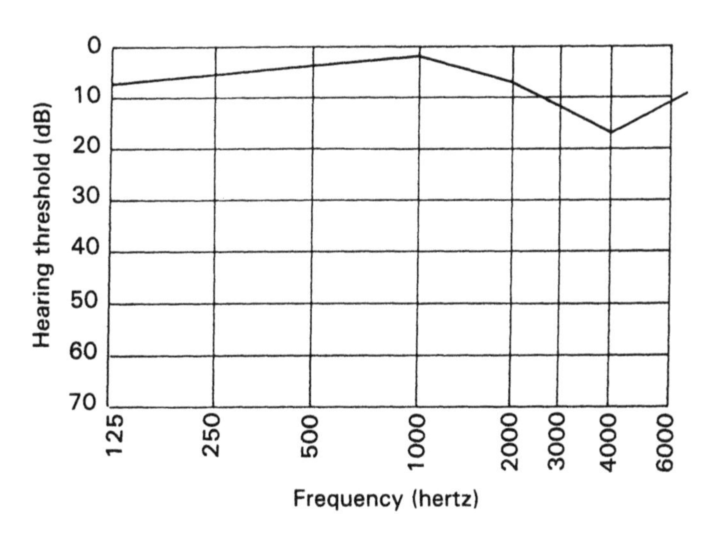

Figure 6.1. Typical audiogram of young person (Fraser, 1989)

Table 6.9. Classes of hearing impairment (after Subcommittee on Noise of the Committee on Conservation of Hearing of the American Academy of Ophthalmology and Otolaryngology).

Class	Severity	Average hearing in better ear (500, 1000, 2000 Hz)	Interference
A	Not significant	<25 dB	No significant difficulty with faint speech
B	Slight	25–40	Difficulty only with faint speech
C	Mild	40–55	Frequent difficulty with normal speech
D	Marked	55–70	Frequent difficulty with loud speech
E	Severe	70–90	Can understand only shouted or amplified speech
F	Extreme	90+	Usually cannot understand even amplified speech

1000, 2000 and 3000 Hz in the under 40 age group, at 35 dB in the 40–50 age group, and at 45 dB in the over 60 age group.

The Subcommittee on Noise of the Committee on the Conservation of Hearing of the American Academy of Ophthalmology and Otolaryngology has outlined various classes of impairment as indicated in Table 6.9.

Lung function

Pulmonary function testing is not a routine procedure in FCA, although it could be required where there is evidence of pulmonary impairment in situations where this could interfere with the proper performance of a task.

Full pulmonary function testing commonly comprises measurement of the following:

(a) lung volumes, which define the size of the lung;
(b) ventilatory capacity, namely the ability to inhale and exhale, a property dependent on the mechanical characteristics of the chest wall and the lung, that is, its distensibility (compliance), and its resistance to movement;
(c) ventilation minute volumes;
(d) distribution of ventilation and pulmonary blood flow;
(e) diffusion characteristics of the pulmonary membrane.

While comprehensive pulmonary function testing requires the facilities of a pulmonary laboratory, simple screening tests, such as measurement of lung volumes, timed lung volumes and ventilatory capacity can be undertaken where required with a simple spirometer.

A spirometer comprises an empty cylindrical drum floating within a tank of water. An air tube passes through the water into the drum so that when air is blown through the tube into the drum, the drum rises out of the water to a height proportional to the volume of air added during exhalation, and falls again during subsequent inhalation. The drum is connected via a pulley system to a counterbalancing weight. A recording

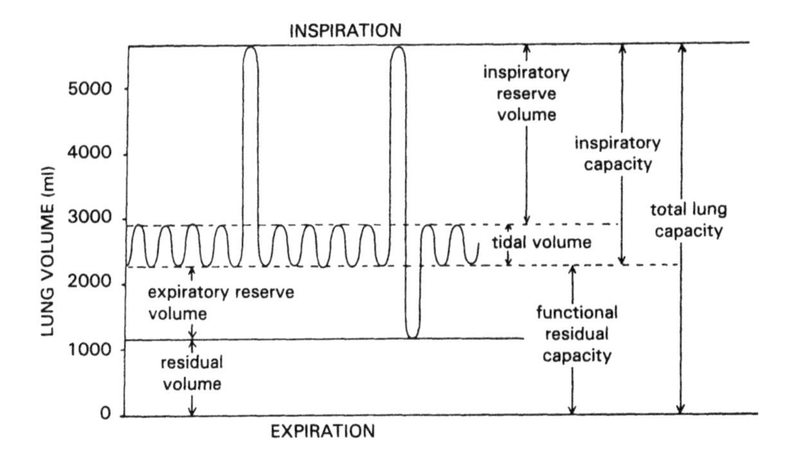

Figure 6.2. Lung volumes and capacities (Fraser, 1989)

device attached to the connecting system records the upward and downward motion of the drum on recording paper fastened around a second drum which rotates at a constant rate. Less cumbersome techniques making use of an electronic device known as a plethysmograph, are in common use in industry and elsewhere for the same purpose.

By breathing, or blowing air in a standardized manner into the spirometer, various static lung volumes and capacities can be defined, that is, those that represent the actual volumes of air contained within the lungs. Measurement of these volumes and capacities, and also various other dynamic volumes, although not normally required for employment purposes, become important in determining the effects and relative disabilities incurred in diseases of the lungs, and particularly in evaluating the effects of industrial dust diseases. The volumes and capacities, as defined from a spirometer, are illustrated in Figure 6.2, and defined as follows (Fraser, 1989):

Tidal volume (TV): The air moved in and out of the lungs during each successive breathing cycle.
Functional residual capacity (FRC): The air left in the lung following completion of a normal expiration.
Expiratory reserve volume (ERV): The air that can be expired by a forced maximum expiration after completion of a normal expiration.
Residual volume (RV): The air remaining in the lung after a forced maximal expiration.
Inspiratory capacity (IC): The air that can be inspired above the level of the functional residual capacity (i.e. after completion of a normal expiration).
Vital capacity (VC): The maximum volume that can be expelled after a full inspiration.
Total lung capacity (TLC): The sum of the functional residual capacity and the inspiratory capacity.

Certain dynamic volumes, that is, volumes of air that can be breathed per unit time, can also be measured, namely:

Forced expiratory volume (FEV): This is the maximum volume that can be exhaled over a period of 1 s. It is expressed as a percentage of the VC. In a normal 20-year-old man it is about 80% of the VC. In obstructive lung disease expiration is interfered with more than inspiration. The flow is limited by the efficiency of the expiratory muscles and by the extent of increase in airway resistance. Sometimes durations of other than 1 s are used.

Maximum voluntary ventilation (MVV): The MVV is sometimes referred to as the 'maximum breathing capacity' (MBC). It defines the maximum volume of air that can be breathed in a 15-s period. The normal level for a 25-year-old male is about 100–180 l, and for a 25-year-old female about 70–130 l.

Normally, routine pulmonary function testing as a precursor to employment would be neither necessary nor desirable. If there is evidence in a candidate of potential lung disturbance which could interfere with the satisfactory performance of his/her task, or if a worker were expected to work in an environment where, in spite of all precautions, there was a potential for release of a chemical constituting a hazard to the lung such as, for example, isocyanate, then preliminary screening would be considered desirable. Such screening might constitute measurement of TV, ERV and IC which can easily be measured by simple spirometry. Measurement of RV, and subsequent calculation of FRC and TLC requires a more complex technique. It is not uncommon also to measure the FEV and the MVV.

Since the volume of air measured by, or calculated from a spirometer is influenced by the ambient temperature, pressure and humidity, it is necessary to bring all volumes to a standard level. This is accomplished by referring all temperatures to degrees absolute, and water vapour pressures to a standard 760 mm Hg, as follows:

$$V(\text{BTPS}) = V \times \frac{273 + t(\text{B})}{273 + t(\text{spir.})} \times \frac{P(\text{b}) - P(\text{water vapour})}{P(\text{b}) - 47}$$

where V(BTPS) = volume at body temperature and ambient pressure, saturated with water vapour (ml); 273 = absolute temperature (°C); t(B) = body temperature (37°C); t(spir.) = spirometer temperature (°C); P(b) = barometer pressure (mm Hg); P(water vapour) = ambient vapour pressure (mm Hg); and 47 = lung vapour pressure at body temperature.

Total concept of FCA

The total concept of FCA is illustrated in the model presented in Figure 6.3. The model demonstrates, as has been emphasized in the course of this chapter, that the requirements of a physical examination for evaluation of fitness to work are job oriented and determined by the demands of the work and the working environment, as modified by the nature of the occupational and general medical histories. The physical examination may be extended if necessary by relevant job-oriented clinical and laboratory testing where required, and also by a work capacity assessment to the extent deemed necessary. The objective of course, is to ensure that the subject can

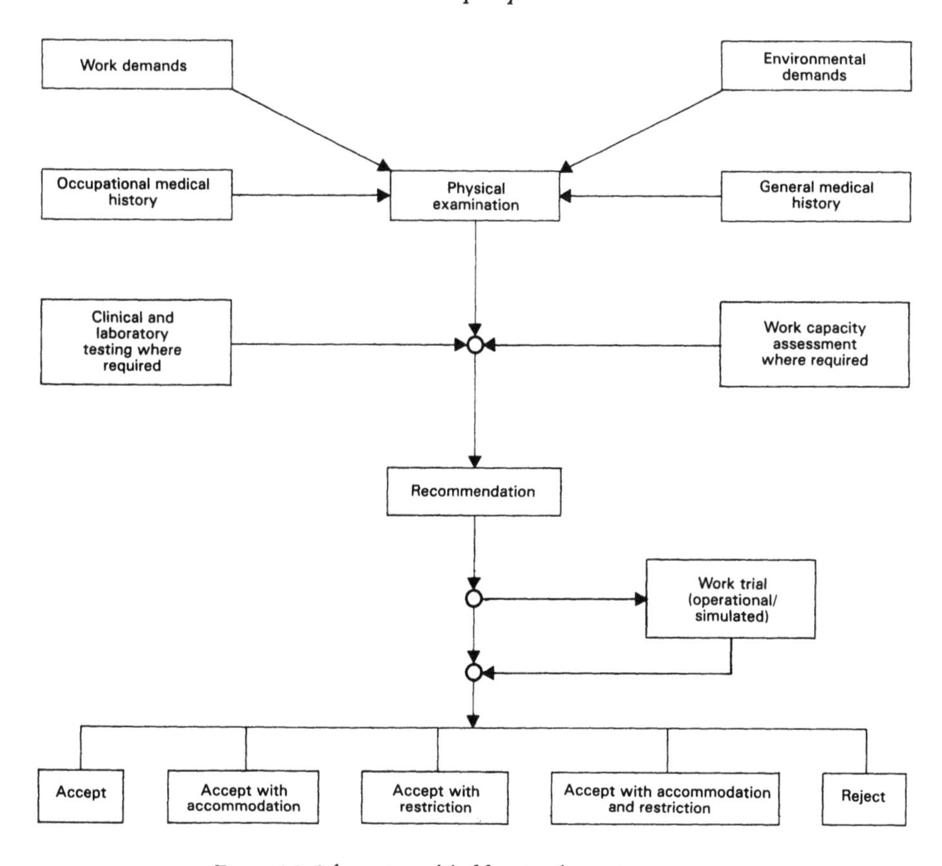

Figure 6.3. Schematic model of functional capacity assessment

undertake the essential demands of the job. On completion of the assessment recommendations are made to the employer. Recommendations may include the need to conduct a controlled work trial, where this is deemed desirable, either under supervision on the job, or in a simulated work environment with simulated work tasks. Work trials will be considered in the next chapter. If the applicant is found to be capable of meeting the requirements of the job, he/she should be considered acceptable; if not, he/she may be accepted along with recommendations for accommodation to his/her impairment in the work place, or with working restrictions, or both. Accommodations and restrictions are also considered in the next chapter. If the candidate is considered to be unacceptable even with restrictions and/or reasonable accomodations, he/she is then rejected.

Chapter 7
Accommodations, restrictions and the handicapped

As noted in the previous chapter, the object of a functional capacity evaluation is to determine whether a job applicant is able to do the essential duties of a job, and the end result is a recommendation on his/her unrestricted acceptance, acceptance with restrictions, acceptance with accommodations, or acceptance with accommodations and restrictions.

Traditional approach

Traditionally it has been the practice that if a candidate was not fully fit for any type of job then he or she was either rejected from employment or permitted to be employed with specific restrictions commonly defined (although sometimes somewhat vaguely) by the examining physician, as outlined in Chapter 2. These restrictions might be general, as in 'fit for light work', or more specific as in 'may not lift more than 10 kg, once per hour'. The possibility of modifying duties, or even more so, of modifying a workplace, to accommodate the needs of a temporarily or permanently handicapped worker was not considered.

Employment equity

The problem with respect to accommodation then becomes one of ensuring employment equity. Employment equity addresses the situation of persons who historically may have been the victims of intentional or inadvertent employment discrimination by reason of impairment or handicap. Its purpose is to achieve equality in the workplace so that no person can be denied employment opportunities or benefits for reasons unrelated to ability, and, in the fulfillment of that goal, to correct the conditions of disadvantage in employment experienced by those persons. Employment equity means more than treating persons in the same way; it may also require special measures in the accommodation of differences, so long as it can be shown that the needs of the person or group can be accommodated without undue hardship. The object is to ensure that a person with a disability receives equal treatment if he/she is capable of performing or fulfilling the essential duties that accompany the exercise of his or her rights. In some circumstances the nature or degree of a person's disability may preclude him/her from being able to perform the essential duties. A person however should not be found incapable of performing these essential duties unless an effort has been made to accommodate his/her needs.

Requirements for accommodation

The requirements for accommodation derive from numerous different laws, regulations and guidelines promulgated by various legislative bodies in the industrial world, pertaining to the rights of the worker and specifically to the rights of the handicapped worker. It is not feasible to outline all of them. Some legal considerations are examined in Chapter 9.

In the USA in particular, which serves to illustrate the situation although the law is not applicable elsewhere, the requirements derive from a series of legislative enactments. Those that have had the greatest impact are the regulations developed to implement Section 504 of the Rehabilitation Act of 1973 which require that 'reasonable accommodation to the needs of handicapped persons' be provided.

As a result of the laws passed under these various jurisdictions several human rights guidelines have been developed to assist companies in implementing the necessary accommodations. In particular, a comprehensive set of guidelines has been developed under the Ontario Human Rights Code (Human Rights Commission, 1981) from which some of the following material is taken.

Individualization as the key to accommodation

The essence of accommodation is individualization. The first step is to determine what is considered essential and what is not. If the person cannot perform the essential duties, accommodation must be explored that will allow him or her to do so. None-essential duties can be re-assigned if necessary or, where feasible, an alternative method can be developed for fulfilling those duties. Accommodation of needs includes, for example, making buildings and transportation accessible, making print information available in alternative formats such as type, or braille, translating auditory information into visual or tactile modes for persons with a hearing impairment, adapting equipment or providing special devices or supports so that the person with a disability will be able to function independently, altering the ways in which tasks are accomplished to allow for a person's disability, and so on. These accommodations should be made in a manner which recognizes the privacy, confidentiality, comfort, autonomy and self-esteem of persons with disabilities, thereby maximizing their integration into society. It is recognized that in some circumstances the best way to ensure the dignity of persons with disabilities may be to provide them with separate or specialized services. This approach may not be acceptable, however, unless integrated treatment would pose undue hardship. The emphasis is on provision of accommodation which is most compatible with the dignity of the disabled person.

Hardship

The onus of proving undue hardship lies with the person responsible for providing accommodation. At the same time, the person who requests the accommodation has a responsibility to communicate his/her needs in sufficient detail and to co-operate in consultations to enable the person responsible for accommodation to respond to the request. There should be objective evidence for determining the financial costs, the effects of the projected costs on the enterprise responsible for accommodation, and for determining the seriousness of the health or safety risk.

There are two types of situation in which the requirement to accommodate may arise. One situation is where a person with a disability is a member of a group against which discrimination is alleged, and that person requests accommodation which would also accommodate the needs of the whole group. The other situation is where an individual requires accommodation for his or her own needs. Where accommodation of the needs of the group would cause undue hardship to the company, but the individual could still be accommodated without undue hardship, then the individual accommodation should be made.

In assessment of hardship, the emphasis is on the *current* abilities of the person and the risks of the *current* situation, rather than what may arise in the future, and indeed the unpredictable nature and extent of future disability should not be used as a basis for assessment of the needs for the present.

It is common practice that cost and health and safety factors are considered to be the prime determinants of hardship. In particular, business inconvenience or undue interference with the enterprise is not considered significant in determining hardship, nor is customer or other third party preference. Equally, in a few jurisdictions, the presence and content of a collective agreement or other contractual arrangement is not permitted to act as a bar to provision of the kinds of accommodation an employee with a disability might require. It is considered to be the joint responsibility of the employer and the union to determine a solution to any conflict with the collective agreement. If a solution cannot be reached, the employer must make the accommodation in spite of the agreement.

Cost

Undue hardship can be shown to exist if the financial costs demonstrably attributable to the accommodation of the needs of the person or group would alter the essential nature or would substantially affect the viability of the enterprise responsible for the accommodation. The costs so demonstrated should be quantifiable and clearly related to the accommodation. The type of costs would include the capital and operating costs, as well as the cost of additional staff time beyond what could be accomplished through restructuring existing resources and job descriptions to provide appropriate assistance to the person with a disability. Other costs might include increased insurance costs or sickness benefits. Additional staff time might include the cost of an assistant or a personal attendant to assist the person with a disability to do his or her job or to use the service or facility provided.

For determining whether a financial cost would alter or affect the viability of the enterprise attention should be given to the possible recovery of costs by way of grants, subsidies or loans from government or other sources, as well as the distribution of costs throughout the operation, amortization or depreciation of capital costs, cost savings by such approaches as tax deduction, improvement in productivity, efficiency or effectiveness, potential increase in resale value of property, increase in labour pool, customers or tenants.

Health and safety

Undue hardship can be shown to exist where the degree of risk after an accommodation has been made or suggested outweighs the benefits of enhancing employment equality for the disabled or impaired person. This occurs particularly where a proposed accommodation creates a potential conflict with a health or safety requirement, whether in law, regulation or in some established practice or procedure. Where the effect of such a requirement is to exclude the person with a disability from the workplace or service, it may be necessary to modify or waive the health or safety requirement. Whether this action will create undue hardship or not depends upon whether the remaining degree of risk outweighs the benefit of enhancing equality.

In determining if an obligation to modify or waive a health or safety requirement creates a significant risk to any person, disabled or co-worker, consideration should be given to the seriousness of the potential risk to others, the types of risks tolerated within society as a whole, the other types of risks which the impaired or disabled person is assuming within the enterprise, and his/her willingness to assume the risk under consideration in circumstances where the risk is strictly personal. The seriousness, in turn, is determined by assessing the nature of the risk, its severity, the probability of occurrence and its scope.

Application of accommodations

Work restriction

The classical approach to accommodation has been that of work restriction, and this is still by far the commonest, simplest and most cost-effective, and where appropriate, the most useful. Restrictions may apply for example to the type of work the employee is called on to do, such as climbing, lifting and driving. They may apply to the duration or distribution of work, such as prohibition of shift work, limitation to daytime hours, part day work, extended rest periods and so on. They may restrict the worker to work only in jobs shared with other workers. They may prohibit the employee from working with certain materials, such as solvents and so on. The restrictions may be temporary, such as during some part of pregnancy, or on return from illness or injury, or they may be permanent because of a congenital disorder or a disability from injury. For whatever reason they are applied, the intent is to reduce the chance of aggravating a condition or injury during employment or to assist the employee in returning to full working capacity. The major proviso in application is that each situation be examined on its merits with careful consideration to the nature of any limitation, and that the most appropriate restriction be applied. It is often desirable to record the restrictions on a specific form prepared by a health professional for distribution to the human resources office and the supervisor, and on which are recorded the specific nature of the restrictions and the duration for which they apply before reconsideration. General blanket restrictions should be avoided.

Job restructuring

While work restrictions apply to the person, job restructuring, as the name would imply, applies to the nature of the job. In its simplest form it entails removal of such non-essential parts of a job that are beyond the capability of a specific disabled person. This, in turn, requires a detailed analysis of the job with identification of the discrete mental and physical tasks involved, either by direct study of the job or by reference to a previously determined physical abilities analysis. Perusal of the analysis, along with discussion with the job incumbent and/or supervisor can elicit those parts of the job that are considered essential and those which can be relegated or even eliminated. For example, an electronic assembly worker susceptible to tendonitis may change places with a non-susceptible worker on the assembly line where the particular actions causing the tendonitis do not occur.

Job or site modification

Job or site modification requires a change in the method of accomplishing the task. Depending on the extent of the change, such a process can be costly, although often it may involve only some simple change of working height so that a wheelchair can fit beneath the work surface. The analysis in this situation should take into account the extent to which the employee can perform the job without special equipment and to what extent the performance is likely to be improved with special equipment, and whether such modification will eliminate or minimize any possible aggravation of the condition under consideration. Nylander and Carmean (1984) list some modifications which have been shown to be successful, including the following:

— acquisition of a tape recorder or dictation machine for an employee unable to write;
— provision of telephone earphones for an arm/hand impaired receptionist;
— installation of oversized touchtone buttons on a telephone for a person with poor motor discrimination;
— provision of a bookstand for easier reading;
— acquisition of visual indicators to replace or co-exist with bell warning systems for the hearing impaired;
— installation of high-contrast furniture for a visually impaired worker;
— overhauling of a ventilation system to reduce smoke, dust and fumes;
— relocation of equipment controls from one side to the other, or change from hand to foot (or vice-versa) to encompass arm or leg disability;
— installation of telephone dialling devices;
— expanded use of word processors for handicapped employees;
— use of computer mediated and enlarged visual presentations for visually handicapped persons.

Support services

Support services involve the use of a third party to facilitate the work of the handicapped person but not to perform the actual essential duties of the job. The service might include someone to guide the worker to the job, or even to bring him/her to the place of employment, to assist in a toilet, or perhaps to communicate in sign language where appropriate. The cost of these services can be high, and in a small company might be construed as causing undue hardship to the company.

Barrier removal

There are two forms of barrier: architectural barriers or structures, and institutional barriers or tests and procedures. Removal of architectural barriers is the most obvious, and often the most effective form of accommodation, although again it can be relatively expensive. Its benefits apply chiefly to the person in a wheelchair, and essentially it comprises the provision of access to all necessary areas in the workplace. It may involve the widening of passages, the provision of ramps, the relocation of signal buttons for use from a wheelchair, the modification of lavatory stalls for use by a wheelchair occupant and so on.

Institutional barriers can be as limiting as architectural ones. Regulations in various jurisdictions require that an employer may not use any employment test, selection criterion, or policy, that screens out a handicapped person from consideration for employment unless the test, selection criteria, or policy as used by the recipient is shown to be directly related to the essential functions of the position in question, and alternative job-related tests, criteria or policies that do not screen out as many handicapped persons, are shown to be available.

Testing of the handicapped

In a definitive paper outlining the problems involved in testing handicapped persons for employment, from which some of the following material is derived, Nester (1984) has discussed various approaches that can be made to testing. She quotes a study by the US Civil Service Commission (1956) in which the validity of a test was studied simultaneously for handicapped and non-handicapped persons. The study elucidated many of the critical issues in the competitive selection of handicapped workers, such as:

- the need to use tests for the blind that are equivalent to those for the sighted so that blind and sighted applicants can be placed on the same employment register of eligible applicants;
- acceptance of the goal of 'practical equivalence' of test scores (through criterion relation) since exact comparability cannot be achieved;
- the relevance of modification of work sites and re-engineering of jobs by transferring tasks that require sight to sighted workers and giving blind workers tasks that do not require sight;
- the existence of *de facto* job accommodation to the abilities of each worker, both blind and sighted;

— the possibility that blind and sighted workers have different ways of doing the same job, and consequently, the possibility that blind and sighted workers might require different abilities to do the same job;
— the question of who should rate the requirements of a blind person's job in cases where there are no blind incumbents.

She notes that although these issues were raised only with respect to blind workers, they are equally applicable to other handicapping conditions.

Testing of the visually impaired

Legal blindness exists where there is a visual acuity with corrective lenses of 20/200 (6/60) or less. About one-third to one-half of the legally blind who may be asked to take written tests have sufficient vision to read large print or to read regular print with a magnifier. Some may have a limited visual field which slows the act of reading. It might be noted also that persons blind from birth may not have concepts of colour, while those who become blind as teenagers or adults may not develop the capacity to read braille.

Many tests have been developed or adapted for testing of the visually impaired. A listing of these is given in Scholl and Schnur (1976). No single method is practical for all purposes. The most common approaches are auditory, by cassette tape or live reader; however, where feasible most people tend to chose print copy rather than an auditory method. All of the methods used are intrinsically slower for the subject than is regular print for the sighted reader. Hence extra time is required for the test and time limitations as a means of competition are inappropriate. The time requirements as a ratio of the time taken by visually impaired subjects for visual tasks to the time taken by 300 normally sighted subjects is shown in Table 7.1.

Verbal tests are considered to be appropriate provided they avoid concepts or phenomena unfamiliar to persons born blind or who became blind in early life.

Table 7.1. Comfortable time limit for visually handicapped competitors as a ratio of the time limit for non-handicapped persons (numbers in parentheses indicate the number of applicants who took the test in that medium) (Nester, 1984).

	Medium			
Test	Large print	Braille	Cassette tape	Reader
Reading comprehension	2·0	3·0	2·3	2·6
	(85)	(88)	(29)	(35)
Vocabulary comprehension	1·7	2·1	2·0	2·4
	(87)	(92)	(29)	(35)
Letter series	3·4	4·1	5·0	3·9
	(103)	(99)	(13)	(25)
Inference	2·4	3·3	2·9	2·6
	(89)	(91)	(25)	(36)
Arithmetic reasoning computation	5·0	7·1	6·9	7·6
	(93)	(84)	(10)	(34)

Examples are colour, shape, visual texture and certain geographical concepts (Bauman and Hayes, 1951).

Testing of the hearing impaired

The age at which deafness occurs is the most important variable affecting the testing of the hearing impaired. Those who become deaf before they become verbally proficient tend to be limited in their verbal skills, as well as in reading and writing (Myklebust, 1964), although striking exceptions are well known. The poor performance is the result of deprivation of language, however, rather than low intelligence. In a study for the US Civil Service Commission, Stunkel (1957) showed deaf students of university (college) level performed significantly lower on all verbal subtests of a psychological test battery (vocabulary, grammar, reading comprehension, arithmetic reasoning) in comparison with students with normal hearing. The two groups were equal in score on figure analogies, but the deaf students were superior on a letter series subtest.

One of the most important factors in the testing of the hearing impaired is to ensure that they understand the test instructions. If necessary an interpreter using sign language, should be employed to translate the instructions and any verbal questions. The subject should be seated near the interpreter in a well lit room with the light on the interpreter so that his/her face is not in shadow. For verbal questions extra time is desirable in comparison with that required for the normal subject. For non-verbal questions, no extra time is required provided that sufficient time has been allowed for understanding the instructions.

Testing of persons with motor impairment

The term motor impairment is construed to include such conditions as cerebral palsy, multiple sclerosis, paraplegia and quadriplegia. As with other categories of impairment, the nature and severity of the condition determine what test modifications might be needed. For example, a person whose impairment lies strictly in the lower limbs might need no modifications other than an accessible test site. A person with cerebral palsy might require verbal testing, or devices for turning pages and marking answers. Extra time would commonly be necessary.

In 1964 the Council on Occupational Health of the American Medical Association wrote (AMA, 1964):

> Comprehensive and documented studies of the performance of the handicapped have repeatedly shown excellent job performance, as well as less absenteeism and better safety records than in comparable groups of able-bodied workers. In most circumstances, such employment does not lead to increased workmen's compensation costs.
>
> The principle of evaluating ability, rather than disability of a potential employee deserves continued emphasis. Strict placement requirements are unavoidable for certain jobs, but if the type of work permits, the handicapped individual should receive equal consideration with any other worker.

The same principles apply nearly 30 years later.

Chapter 8
Job matching

Theory of job matching

The ultimate object in conducting a physical demands analysis and a functional capacity assessment is to provide a structure on which to match the capacity of the worker with the demands of the job.

The process of job matching depends on recognition of the relationship between worker capacity and job demand. The matching process can be viewed as the establishment of a value for the ratio of worker capacity (WC) to job demand (JD). Three forms of this relationship can be defined as follows:

$$WC = JD$$
$$WC > JD$$
$$WC < JD$$

Ideally, for the most feasible match,

$$WC/JD = 1$$

that is, where the work capacity equals the job demand, the working situation should be ideal. Where the work capacity is less than the job demand, or for that matter greater than the job demand, then the working situation is not ideal.

To make this judgment, of course, it is necessary to bring both working capacity and job demand on to the same scale to the extent feasible, a feat that is possible only to a limited degree. In fact it is common practice to make that judgment on a more or less subjective basis without using formal scaling systems in what might be termed the qualitative approach to job matching. While such an approach can be effective, it lacks the objectivity that can be achieved by a more quantitative approach embodying direct comparison of measurable qualities or characteristics common to both work capacity and job demand. Both qualitative and quantitative approaches will be considered in the following sections.

Practice of job matching

The fundamental concept of job matching has been noted in Chapter 1, where it was stated that some of the earliest approaches to classification of work and job demand could be found in the military services. In fact, by the end of World War II a degree of sophistication had been developed in that classification process as it became necessary, particularly in the naval and air forces, to classify recruits and serving personnel in groups according to their apparent suitability for battle, for support or for special purposes such as flying and sea duty. It was also necessary to develop categories defining temporary suitability, or suitability for permanent limited duties.

This type of evaluation, and others of a similar nature, required a comprehensive medical, personal and family history, which was not specifically job oriented, along with physical and laboratory examination of the subject with the object of defining:

(a) stable impairments, that is, those that cause some limitations but are not likely to progress or be worsened by the job (or military service), and do not interfere with general working capacity;
(b) impairments, stable or unstable, that are likely to be worsened by work exposure or military duty, and may interfere with general working capacity;
(c) progressive conditions which may result in increasing limitation;
(d) conditions of intermittent impairment which may occasionally result in incapacity (e.g. epilepsy).

In military services during and after World War II, however, the codes and descriptive codicils became quite complex, particularly in the air forces. Concomitant with the medical classification of personnel there developed a parallel system for defining the requirements of the job. For example, in the Royal Canadian Air Force, which had a highly refined system, the term A1B defined a certain high level of physical fitness; it also defined the requirements for flying as a pilot, and fitness for full ground duties. Similarly, the term A3B defined a different level, and allowed flying as a navigator. AtB meant temporarily unfit for flying, but fit for all ground duties, and in even greater complexity, ApBt, for example, meant permanently unfit for flying, and temporarily unfit for ground duties, in a veritable lexicon of codes. On the basis of the history and examination, however, and in the light of the general physical demands outlined by operating military authorities, it then became possible to assign serving personnel or recruits to broad categories of potential usefulness.

As the need grew for more precision in definition, a still more sophisticated coding system came into effect, originating in the British military services, but modified further in Canada. The model that was used went by the acronym of PULHEMS, where each letter of the acronym designated a particular body region or function, namely, *P*hysical, *U*pper limbs, *L*ower limbs, *H*earing, *E*yes, *M*entality, and *S*tability, and where each category was graded on a numerical scale, commonly from a high of 1 to a low of 5.

Each grade in the system was defined in terms of the standard necessary to achieve a particular rating such that, on completion of the required assessment, a subject could be assigned a numerical profile, for example 1122122, where each number referred to a grade in the PULHEMS system and thereby could define the person's fitness in fairly precise terms. Unfortunately the system became misused by people who failed to appreciate its elegant simplicity and began to distort the ratings to mean such broad and indefinable categories as 'fit for light duty', 'fit for shore duty only' and so on, rather than leaving it as a descriptive record of functional capacity and physical demand.

The military normally deal with a healthy young population. The basic objects of military physical evaluation are firstly to define a population of potential recruits who are unsuitable for military service and reject them accordingly, and secondly to categorize the remainder into categories that can define the service they can perform from the demands that are made. In industry, on the other hand, one is not necessarily dealing with a young healthy population who can be rejected as authorities see fit. The object of the process should be not only to define those suitable for all needs, but also to determine the extent to which the less fit, the impaired, the disabled and the handicapped can participate in meaningful employment, and to match their capacities with the essential demands of specified jobs.

As noted in Chapter 1, an early approach to this type of matching was undertaken by the US War Commission in the mid 1940s, based on comparing job requirements, defined in broad terms by a short abstract, against certain arbitrarily defined human performance abilities. Their work, however, was characterized by first defining certain predetermined abilities such as standing, sitting, crouching, walking, climbing, throwing and so on, and then determining to what extent a candidate could achieve these demands.

Clearly, it would be still better to approach the analysis from the other end, as it were, to first define the demands and then see to what extent a given candidate can achieve them. This, of course is the principle of physical demands analysis (PDA), discussed in detail earlier. Having completed a PDA and an associated functional capacity assessment (FCA) it then becomes possible to match one with the other in the job matching process.

The qualitative approach to job matching

There are no hard and fast rules as to how this matching should be done. In the absence of formal quantitative or quasi-quantitative job matching techniques, it is commonly the function of human resources personnel to assemble available facts from health professionals and job analysts and make the judgment accordingly. Indeed, in large organizations there may be a placement officer whose primary function this may be. Depending on the nature of the methods used the judgment may be made in a subjective, non-systematic manner or in a more systemized quasi-quantitative manner. In either case, however, the judgment must be capable of standing up to legal, humanitarian or other challenge.

Ideally, and particularly when being done in a qualitative manner, a job matching programme is conducted by human resources personnel knowledgeable about job activities and human limitations, in co-operation with health professionals, job analysts, supervisory personnel and even incumbents.

Any approach, of course, even if only qualitative, necessitates a careful, thorough and specific PDA of the task, job or jobs concerned to determine what are the specific requirements of the position under consideration. Only then is it possible to determine with reasonable assurance from a corresponding FCA of a potential incumbent whether in fact there is a likely match between the worker capacity and the job demand. Certain broad generalizations may be outlined, however (Abt Associates, 1984). Firstly, the process of job matching has to be able to stand the test of practicality. In particular it should be a cost-effective alternative to current practice, relatively quick to apply, involving minimal paper work and amenable to computerization. If standardized tests are used they should be defined, or the assessor should state how conclusions have been reached, for example by observation, deduction and qualitative appraisal if quantitative methods are not available.

Regardless of method however, if the matching process includes a limitation with or without accommodation it should be time-limited, that is, the profile should include a date beyond which the results are invalid and reassessment is required, since once diagnosed or labelled a worker may retain a diagnostic or restricted rating. Illnesses and diseases, however, can be cured or controlled, or for that matter they can worsen. New skills can be learned and compensation for impairment can be developed. It is therefore necessary to ensure that a label once applied does not become permanent without proper consideration and reconsideration.

Thus, in practice, the placement officer reviews the job demands, commonly presented in narrative or partial narrative form, to obtain a greater or lesser understanding of the requirements of the job. With these in mind he reviews the capacities and limitations of the worker as presented by the medical documentation and attempts to seek a match. Unfortunately, unless there has been some attempt at formal FCA, medical documentation is usually presented in much less useful detail than job analysis which, if for no other reason, has generally been undertaken for pay scaling reasons. Consequently a successful match may be difficult to achieve by this means without more detailed knowledge of both worker and work. Indeed, unwarranted rejection of an apparently unsuitable worker, or equally unwarranted acceptance of an apparently suitable worker could well be the outcome unless resort is taken to a work trial or work simulation.

Job matching should thus be viewed as part of a continuing process. The initial evaluation cannot stand alone as long as there is a possibility of mismatch. Any mismatch or potential mismatch requires further action in the form of repeated, perhaps more detailed assessment, with specifications of other or further aids, accommodations, modifications, rehabilitation, treatment and so on.

Not the least important consideration, which applies equally to qualitative and quantitative approaches to job matching, is the requirement for evaluation and job matching records to be communicable without the use of jargon or excessive technical language, in a manner easily understood by lay persons. It should be recognized that in some, if not many, instances, a wide variety of professionals may contribute to the total process. Each will have and use skills and language peculiar to his/her profession, while in addition job incumbents and supervisors will also tend to contribute technical language. The ultimate users of the information, however, will by lay persons, supervisors and personnel officers, with perhaps limited knowledge of specialized terminology and jargon, except their own. Information, therefore, should should be communicated as much as possible in a standard form of language, determined by mutual discussion, with meanings known to all.

The quantitative approach to job matching

The use of the term quantitative in this connection serves to indicate that the characteristics of both job demands and worker capacity have been defined in measurable or scalable ratings such that one can be directly compared with the other. Several rating attempts have been made in this regard. The methodologies discussed in Chapters 4 and 5 under the headings of 'Physical demands analysis' and 'Physical abilities analysis' lend themselves to a quantitative or quasi-quantitative matching process, but while they are very comprehensive they are also complex. Two simpler but effective methods are presented here, namely the Hanman and the GULHEMP methods.

The Hanman method

Application of the original approach of Hanman (1948) requires the services of three persons, assisted by specialists if necessary. The three persons are, as already noted, firstly, a job analyst who prepares specific information on the physical and environmental demands of jobs; secondly, the occupational health professional who prepares specific information on the fitness and work capacity of the worker; and thirdly, the placement officer, who relates the physical abilities of the worker to the demands of the job and selects the most appropriate job.

As applied either to analysis of job demands or assessment of work capacity, the procedure involves the use of objective units of expression instead of the subjective terms commonly used in qualitative methods. The principal objective unit used is the hour, or part-hour. Then, together with other specific units such as the pound or kilogram, the hour is used to express the safe maximum degree of ability the health professional believes the applicant possesses for a series of standard activities, for example, the ability to lift up to 10 kg intermittently for a total of 1 h per day, to walk throughout that day for a total of 6 h, and to climb stairs intermittently for a total of 30 min per day.

Thus, instead of using subjective terminology such as 'no excessive climbing', the health professional defines what is meant by excessive in terms of time and frequency, for example not to exceed 30 min per day, and not to exceed 20 times per day, whichever is the greater.

A modified version of the Hanman method, based on earlier models developed by Schouppe and Couch (1988), and by Luchterhand and Sydiaha (1966), is shown as Appendix D. The records comprise two forms, namely, a Physical Demands/Physical Capacities Form and an Environmental Conditions Form. The Physical Demands/Physical Capacities Form is used by the job analyst to define the demands of the job, and the same form is used by the medical evaluator to define the physical capacities of the worker.

The Environmental Conditions Form is completed by the job analyst by inserting check marks into the appropriate boxes.

The resulting information with respect to physical demands and working capacities is then compared by the placement officer, who notes any requirements for accommodation, modifications, restrictions or work trials, and uses the information for the assignment of workers or candidates for work.

The GULHEMP method

For his study on employment of the older worker Koyl (1974) developed a rating system based on the previously noted PULHEMS approach. His system goes by the acronym GULHEMP, and defines attributes of the four functioning divisions of the human organism, namely vegetative, locomotive, perceptive and integrative, in terms of *G*eneral physique, *U*pper extremities, *L*ower extremities, *H*earing, *E*yesight, *M*entality and *P*ersonality. He avoided as much as possible the previously noted limitations which had rendered the PULHEMS system inadequate. As he states:

> Many systems have been devised and are in use for various special purposes. Some have been distorted by administrative restriction until they are no longer profiles. Others have been distorted so they are a profile of some arbitrary measurements rather than function. The PULHEMS system exemplified each of these variants . . . Indeed, certain grades represent restrictions on service due to climate etc. Their grading cannot, therefore, be called either functional or a profile. The explanatory pamphlet requires frequent amendment by administrative orders.

In developing his system, then, Koyl (1974) used strictly functional ratings. As a concession he allowed the use of suffixes to a rating, such as R, to indicate that a condition is remedial and may change with time. Other suffixes are used to define specific limitations, as required and defined by the user.

Each component of the scale, that is, G, U, L and so on, is rated in seven levels of effectiveness, where 1 represents the capacities of a very fit person, and 7 means more or less complete incompetence. Each grade level is defined in detail in the appropriate Manual. These grade levels are reproduced in Appendix E and, while everyone may not agree with the categorizations, they represent a systematized approach to classification, and indeed may be open to further modifications by the user.

As with the Hanman approach, the important feature of the GULHEMP system is not merely the use of a systematic, quasi-quantitative and clearly defined scale of values, but the fact that the same GULHEMP scale is used to define the physical demands of the job. In this connection Koyl (1974) emphasizes the need for a team approach in making assessments. In practice, he recommends that the assessment of physical demands be carried out by a committee comprising the physician as chairman, along with a job analyst, a foreman or department manager from the department being studied, a psychologist, as required, a production manager, superintendent, union representative and so on, according to different needs. While many might quarrel with the make-up of the committee, and the fact that the chairman should be a physician, there is no doubt that the principle is sound.

The final assessment of both applicant and task is made in the form of a numerical profile, for example, 2224344, or a graph as seen in Figure 8.1.

This profile, when assigned by a health professional indicates the physical capacities of the worker. For example the first 2 in the profile refers to a general physical

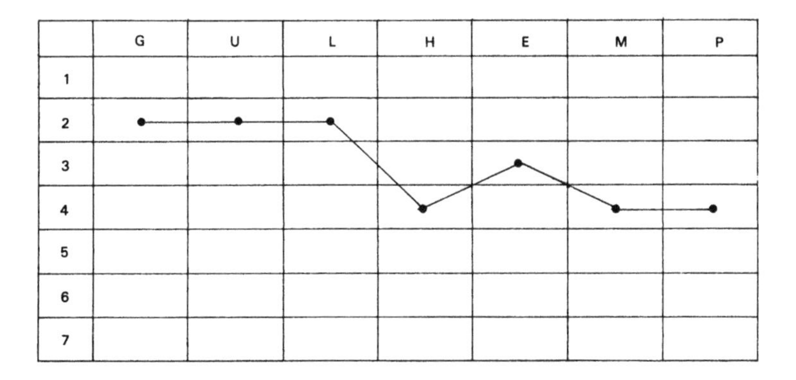

Figure 8.1. Example of GULHEMP profile (Koyl, 1974)

capacity of G2, and is defined in the Manual (see Appendix E) to indicate that the worker is fit for general manual work including incidental or occasional heavy work of the type defined in G1, and that he/she could work in shifts; the second 2 indicates that he/she is fit to lift strongly to and above the shoulder level, to dig, push or drag strongly, and fit for manual work with the upper limbs as is defined in category U2, as well as being fit for incidental or occasional work of the type defined in scale U1; the third 2 indicates that he/she is fit for heavy labour with the lower limbs, as defined in scale L2, and able to stand, run, climb, jump, push incidentally or occasionally, as in L1; the first number 4 indicates that he/she is able to hear well enough to work where severe impairment of hearing is not a handicap (H4), while the number 3 indicates that he/she has the ability to see well enough with one eye to do work not requiring binocular vision (E3); the second 4 indicates that he/she has intelligence sufficient for routine tasks under supervision, could learn related tasks and undertake semi-skilled and unskilled labour, and would be considered a high grade defective with IQ 90–80 (M4), while the third 4 indicates what his/her personality is such that he/she would require encouragement and support, but would be capable of co-operation with others (P4). Such a profile might fit the employment category of dietary porter in a hospital.

Validity of the job matching concept

As noted earlier, the practice of job matching is based on the concept that where worker capacity equals or exceeds job demand there will be a suitable fit between worker and job. This concept was examined by Fraser and Samuel (1979) in both a practical and simulated setting.

In the practical setting the records of 105 workers in an aircraft assembly plant were reviewed. The records were very comprehensive and included full medical and work history data over periods of up to 18 years. Each of the workers selected had worked at one job during his employment period.

In the first study, which involved the assessment of real persons in real work, the data were used to determine the match between pre-employment physical characteristics of the worker and his job, utilizing a Hanman approach. Information was also gathered to determine the extent of absenteeism, and reported injury, as well as diagnostic, treatment and follow-up visits to the medical centre over the period in question.

To test the hypothesis that failure to match worker capacity with job demand would lead to adverse results, the population of 105 workers was divided into three groups by length of service, in each of which, according to the work of Melvian and Maxwell (1974), a different form of resulting behaviour might be expected. These groups were as follows:

Group T1 (probationary): 1 day to 6 months
Group T2 (steady state): 6 months to 7 years
Group T3 (wear and tear, fatigue): 7 years to retirement.

Pre-employment work capacity and job demand profiles were assigned for each worker and each job in each of the three groups. Profiles were determined by the skilled rating of medical and personnel professionals, weighted, summed and averaged to produce a final numerical rating.

The ratio of work capacity (WC) to job demand (JD) was then calculated for each worker. On the basis of this ratio he/she was then assigned to one of the hypothesized conditions, WC = JD, WC > JD, or WC < JD. From personnel and medical records the number of sickness- or injury-related episodes for each person was also determined. Frequency rates for these episodes were calculated for each experience group and predictions made according to the calculated ratio. Results are shown in Table 8.1.

Table 8.1. Frequency rates for episodes of sickness or injury of workers by experience group, and predictions of worker capacity/job demand (WC/JD) ratio (Samuel and Fraser, 1979).

Hypothesis	Employment group	Frequency of episodes	
		Occupational	Non-occupational
WC = JD (ratio = 1 ± 0·05)	1	0·2	0·3
	2	0·1	0·4
	3	0·3	0·5
WC > JD (ratio > 1·05)	1	3·0	3·3
	2	3·1	3·4
	3	3·2	3·5
WC < JD (ratio < 0·95)	1	3·5	3·0
	2	3·6	3·4
	3	3·7	3·1

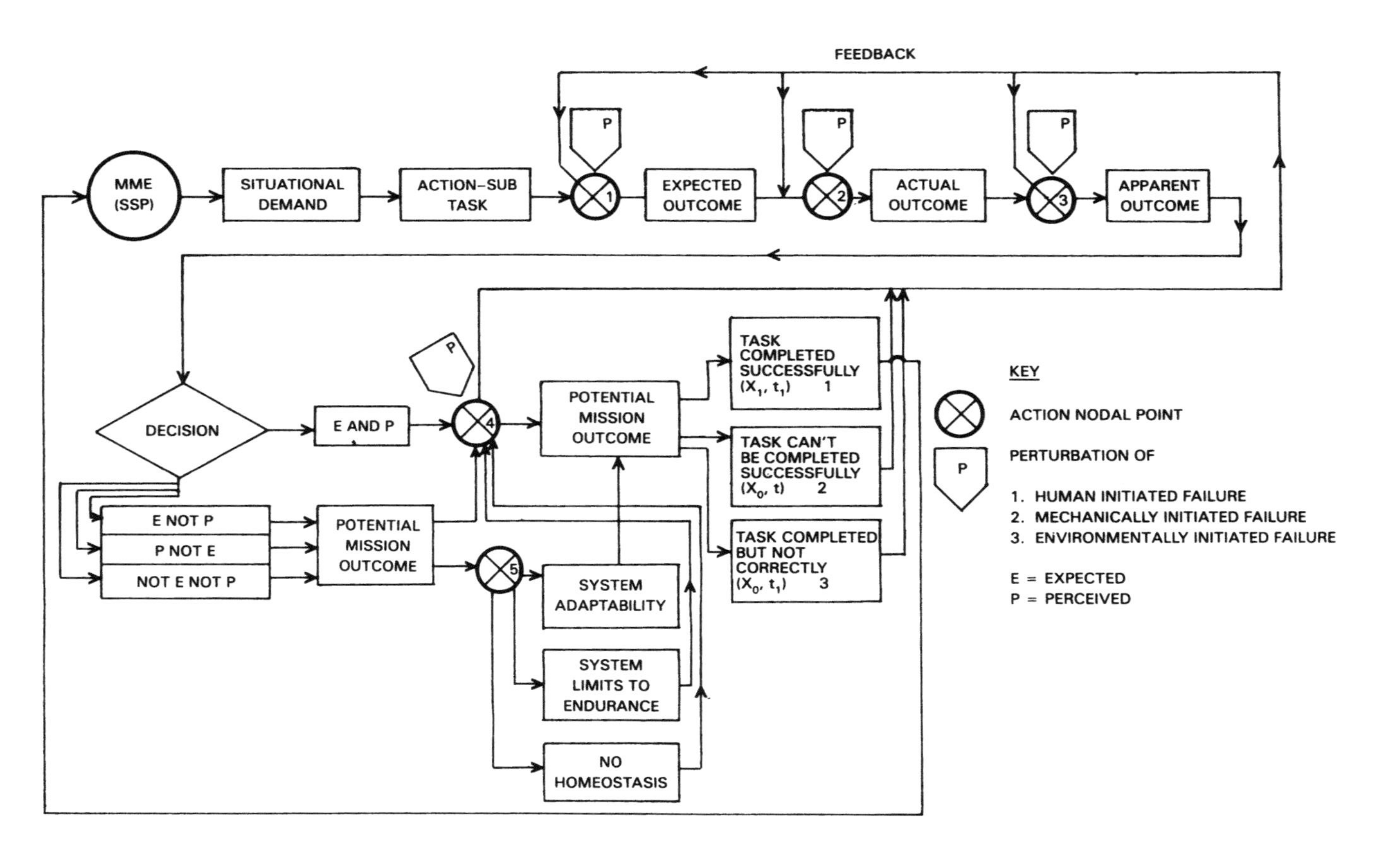

Figure 8.2. Model of human task performance (Samuel and Fraser, 1979)

Table 8.2. Number of sickness and accident episodes when worker capacity (WC) equals job demand (JD) and WC is not equal to JD by pre-employment profile simulation, for 10 000 workers through 100 000 cycles per level (Samuel and Fraser, 1979).

	Employment T1			Employment T2			Employment T3		
WC rating	WC = JD	WC < JD	WC > JD	WC = JD	WC < JD	WC > JD	WC = JD	WC < JD	WC > JD
0	49	—	70	24	—	20	12	—	10
1	51	70	72	23	17	18	12	11	12
2	52	61	63	11	12	22	7	12	14
3	43	61	63	21	24	17	10	17	8
4	45	56	59	18	27	21	9	9	11
5	57	75	62	20	17	20	9	7	12
6	53	51	78	18	17	17	12	9	9
7	51	70	75	10	18	19	9	10	12
8	49	51	80	14	19	20	7	12	12
9	40	43	53	17	18	21	7	10	10
10	35	39	46	12	13	24	6	12	15

The results show that where the pre-employment work capacity meets the requirements demanded by the job the frequency of the occupational and non-occupational injury or disease is markedly less than in the situation where the worker capacity was either less or greater than the job demanded. No significant difference was generated, however, by duration of employment.

In a second study a simulation of the accident- and illness-related behaviour of workers in the same three levels of employment was also conducted, using the technique of general purpose simulation (GPSS) as applied by Gordon (1975) to discrete system simulation.

The approach to the simulation is outlined in Figure 8.2 in a model (Samuel and Fraser, 1979), the original form of which was developed by Fraser (1977). The model is used to outline the course of events in the process of pursuing a task.

Three outcomes are defined for the system operation, namely, successful completion of the mission or task, failure to complete the mission, and incorrect competion of the mission. The model demonstrates that when a demand is placed on the system an action will occur, commonly in the form of a subtask. From the action will come an expected outcome, which, because of system perturbations, may or may not be the same as the apparent outcome. The perturbations can take three forms:

(i) human initiated failure,
(ii) mechanically initiated failure,
(iii) environemntally initiated failure.

The potential outcome, which is also subject to the effect of perturbation, depends on the extent to which the apparent outcome is expected and perceived.

Using this model, a study was made of the probability of mission (task) completion by 10 000 computer-simulated workers cycled through 100 000 computer-simulated tasks for each of the previously defined employment levels (T1, T2, T3).

Table 8.2 presents the number of sickness and accident episodes at different levels of employment in this simulation. The data are presented for each of the three hypothesized conditions at pre-employment WC profile ratings from 1 to 10. Table 8.2 demonstrates that where job demand equals worker capacity in the simulated probationary and steady-state conditions of employment, there are fewer episodes of sickness and accidents or accident-related events.

While these studies are not conclusive, they present some evidence to suggest that the job matching process is effective not only in assisting the provision of suitable employment for the impaired or handicapped, but also in reducing the potential incidence of both occupational and non-occupational accidents or injuries.

Chapter 9
Some legal considerations

In today's society where increasing consideration is being given to human rights, the rights of the disabled and handicapped are becoming more significant. Throughout the industrial world legislation has been enacted, and to a greater or lesser extent enforced, to ensure that the presence of a disability does not give grounds for discrimination either in application for a job or assignment to a job simply because of its existence. A worker may be barred from a job because it is impossible for him/her to do the essential duties of that job by reason of disability, but not simply because a prima-facie impairment exists.

Legislative approaches

It is neither practicable nor profitable to discuss in detail the legislative approaches taken by various nations, states and provinces. Representative examples, notably those with which the author is familiar, will be presented as required. This material, however, should not be construed as being legal authority for any action a potential employer might wish to take. Before implementing any activity with respect to the disabled the reader must become familiar with, and conform with, the national, federal, state or provincial laws, regulations and guidelines that apply to his/her particular jurisdiction.

Internationally, the responsibility for promoting services for the disabled and handicapped is shared among branches of the United Nations, the International Labour Office, and the World Health Organization, with the International Labour Office in particular being responsible for special recommendations and publications (Anon., 1976, 1978).

The earliest action was taken by Great Britain in 1944 by way of the Disabled Persons (Employment) Act, 1944, which established a register of disabled persons, provided a placement service and enforced the employment of a quota of disabled persons by all employers with 20 or more workers. A quota system also operates in Austria, Belgium, Brazil, France, Germany, Greece, Italy, Japan, Luxembourg and Portugal (Moniz, 1983). Many countries, including the USA and Canada, are required to conform with the written Constitution, and the laws, regulations and guidelines issued by lower levels of government.

One of the most advanced systems is found in Sweden which operates an employment programme for all types of persons handicapped by physical, mental, emotional and/or social conditions. The programme works actively with the Swedish Employment Service in co-operation with hospitals, social services, drug and alcohol services, as well as with penal and corrective institutions, welfare agencies, handicapped organizations and social and sickness funds. Sheltered employment schemes, wage subsidies and tax incentives are also used.

Pertinent concepts

In the USA the original relevant piece of legislation was the Civil Rights Act of 1964, Title VII of which precludes employment discrimination on various specified grounds. Since that time various precedent-setting cases have appeared before the courts from which has developed a body of theory pertaining to discrimination, as well as to the validity of the job analysis procedures used to define jobs suitable or unsuitable for the impaired. Although US case law does not apply elsewhere, the principles elucidated can be of value in considering discrimination in other jurisdictions.

In this connection, practices that discriminate lawfully and those that discriminate unlawfully must be distinguished. Hogan and Quigley (1986) discuss some of the theories originally developed by Schlei and Grossman (1976), pointing out that unlawful discrimination can take the form of (a) disparate treatment, (b) policies or practices that perpetuate past deliberate discrimination, and (c) adverse impact.

The term disparate treatment defines the situation in which the defendant is alleged to have treated individuals differently from the way he or she treated those similarly situated, by reason of race, sex, national origin, religion, age or handicap. The key lies in the analysis of supposed equal treatment, and the primary focus is on the defendant's motivation or intent to discriminate. Although the burden of proof is on the plaintiff to make a prima-facie case, direct proof of intent is not necessary. A preponderance of evidence is sufficient.

The concept of adverse impact theory is dependent on whether the defendant's policies and practices have an adverse impact on a class of people even if there is alleged equal treatment of all persons. The plaintiff has to demonstrate that there is an adverse impact on a protected group. Once this is established, then the defendant has to show some defence indicating that any adverse impact was not the result of unlawful discrimination.

Recognized defences

Four such defences have been defined (Schlei and Grossman, 1976). These are that the selection criteria for employment must be job-related and necessary, that they are a business necessity, that they represent a bona fide occupational qualification, and that they are part of a bona fide seniority or merit system. With respect to job-related criteria the Civil Rights Act of 1964 (section 703 (h)) points out that it is not an unlawful practice for an employer to give and act upon the results of any professionally developed ability test provided that such test is not designed, intended or used to discriminate. This provision of course raises the question of the validity of the testing, a concern which will be dealt with in a later section.

As Hogan and Quigley (1986) note, the business-necessity defence is broader than the job-relatedness concept. Business necessity is not specifically mentioned in the Act. Its application derives from case law. In the case of *Robinson v. Lorillard Corp.*, two qualifying concepts are emphasized: that there must be 'an overriding legitimate business purpose such that the practice is necessary to the safe and efficient operation of the business', and that there must be available 'no acceptable alternative policies or practices which would better accomplish the business purpose advanced, or accomplish it equally well with a lesser differential impact'.

The third defence, a bona fide occupational qualification, is typically used in alleged sex discrimination. To be used successfully the employer must first establish that the classification (e.g. gender) is related to job performance, secondly that it is *necessary* for successful job performance, and thirdly the job performance affected is the essence of the employer's business operation. Hogan and Quigley (1986) note that such job classifications might include restroom attendant, wet nurse and sperm donor!

Content validity

As has been noted on several previous occasions, a careful job analysis lies at the root of employment selection procedures, and indeed many court decisions depend on firstly whether a job analysis was completed, secondly whether it was adequate, and thirdly whether the approach selected was appropriate. US courts have recognized a number of validation techniques that are professionally acceptable, and in particular those outlined in the Uniform Guidelines on Employee Selection Procedures (43 Fed. Reg. 38290–38315, 1978). Any selection procedure, including those involving the measurement of physical abilities, should be developed in accordance with the Uniform Guidelines and Professional standards. It might be noted, however, that the Uniform Guidelines state that any method of job analysis may be used if it provides the information required for specific validation strategies. The courts have considered validation as including the validity of the results and the suitability of the method used. If the job information is considered to be accurate, the next issue addressed is that of degree of evaluator or rater agreement. The courts recommend that data be collected by qualified appropriate respondents chosen either on a random or a systematic basis, to include such persons as job incumbents, first-line supervisors, subject matter experts, or professional job analysts. The agreement between different kinds of job analysis methods is also considered to be important. Differences must be negligible or capable of being rationally reconciled.

In the case of *Chance v. Board of Examiners* (3 Fair Empl. Prac. 673, 1971), quoted by Thompson and Thompson (1982), the court rejected a cursory job analysis that consisted of gathering statements of duties and consulting with job experts. The same court in *Vulcan Society v. Civil Service Commission* (5 Fair Empl. Prac. 1229, 1973) stated that in order for a test to be content valid, the aptitudes and skills required for successful examination performance must be those aptitudes and skills required for successful job performance. Similarly, in *Kirkland v. Department of Correctional Services* (7 Fair Empl. Prac. 694, 1974), the judge stated that the cornerstone in the construction of a content-valid examination is the job analysis. Without such an analysis to single out critical knowledge, skills and abilities required by the job, their importance relative to each other and the level of proficiency demanded as to each attribute, a test constructor is aiming in the dark and can only hope to achieve job relatedness by blind luck. In the light of this viewpoint, Thompson and Thompson (1982) go on to define three criteria for a test to be considered content valid:

(a) the knowledge, skills and abilities tested for must be critical and not peripherally related to successful job performance;
(b) portions of the examination should be accurately weighted to reflect the relative importance to the job of the attributes for which they test; and
(c) the level of difficulty of the examination material should match the level of difficulty of the job.

Still another issue is the extent to which the job analysis is compatible with the strategy selected for its validation. The job tasks and responsibilities must be defined, and the relative frequency, importance and skill level identified in such a manner that the content of the analysis can be validated against a suitable predetermined strategy.

The same principles apply with respect to physical abilities testing. To establish content validity it must be shown that the bahaviour entailed by the test procedures is representative of the job behaviour. Push-ups, pull-ups and so on are seldom representative of the behaviour actually required of a physically demanding job, although tests of that nature may be measures of generalized physical fitness. Courts have, however, accepted tests where they result from a job survey and reflect critical duties or demonstrate practical significance, preferably when they are administered and scored objectively. For example, in *Hardy v. Stumpf* (17 Fair Empl. Prac. Cas (BNA) 471, 1978), quoted by Hogan and Quigley (1986), the court accepted a test of scaling a 6 ft fence even though it disqualified six times as many men as women, since an acceptable job analysis showed a requirement for such an activity. On the other hand, tests designed to simulate real-life conditions on the basis of the experience of some persons were not considered acceptable because the job requirements and the test content were not linked.

In summary, it can be stated that legislation, wherever enacted, and the practices of human rights codes, demand that physical assessment of workers should be related to the essential duties of a job. Commonly enquiries are prohibited with respect to physical or mental conditions prior to the stage of job interview. Enquiries after that stage may be made with respect to the ability of the applicant to perform the essential duties of the job, and after the person has been hired various assessments may be undertaken to determine the physical capacities of that person. Normally, however, any such examinations after hire are restricted to determining information required by the employer.

Conclusions

In this book an attempt has been made to show, initially, that the classical approach to the assessment of fitness for work by way of undirected and often unthinking pre-employment or pre-placement medical examinations, however comprehensive, has often failed to provide the information needed for proper assignment of a worker to a job, and even more so for the best utilization of the impaired or handicapped worker. The adverse results of this failure include a reduced effectiveness of the work-force, with perhaps a higher incidence of injury, sickness and absenteeism than might otherwise be the case, as well as a high level of social cost in maintaining the impaired and handicapped who might otherwise be better employed. The resulting apparent unemployability can lead to still further loss of self-esteem and dignity of these same impaired and handicapped persons. In addition, unthinking and unnecessary medical examinations, not directed strictly to the needs of assessment of work capacity for job demands, can encroach on the privacy and legislated civil rights of the individual worker or job applicant.

It has also been shown that the root cause of the problem lies in the failure to match the specific demands of the job with the specific work capacity of the worker. This matching process requires a careful and comprehensive analysis of the job demands, and a careful job-oriented assessment of the work capacity of the worker before a proper match can be made. An attempt has been made in this book to show how these may be best accomplished so as to encourage the most feasible match.

Appendix A
Physical demands job analysis

This Appendix consists of sample completed forms for a job of dishwasher (DOT — Kitchen Helper 318.687–010), a low-skilled job that requires medium physical demands. The job requires walking, standing, lifting up to 50 lbs, manipulating, reaching, pushing/pulling, twisting the back, climbing, crouching and kneeling. There is a need to move frequently around the workplace, and to some degree outside the building (after Lytel and Botterbusch, 1981).

Summary Form

Employer job title *Dishwasher* DOT title and code *Kitchen helper (318.687-010)*

Supervisor *Mary Swartz* Telephone ______

Company name and address *Howard Johnson's*

Highway 12

St. Paul, Minnesota

Job summary *Washes dishes, silverware, utensils, pots, and pans and cleans the premises and equipment in a restaurant*

No. of employees in same positions at work unit: [1] [2] [3] [X 4] [5] [6] [7] [8] [9 or more]

No. of employees in other positions at work unit: [1] [2] [3] [4] [5] [6] [7] [8] [X 9 or more]

Environmental and social conditions form

1	Inside	8	Mech	14	Sharp	20	Around
2	Outside	9	Electric	15	Floor	21	Alone
3	Heat	10	Hot	16	Elevat	22	Supervision
4	Cold	11	Rad	17	Light	23	Violent
5	Wet	12	Vent	18	Fumes	24	Shifts
6	Noise	13	Mov Ob	19	Others	25	Equipment
7	Vibr						

Task Analysis Form

Most common posture

	26	Walking (or mobility)			
27S	28	Standing	29 Sitting	30S	31 Stooping, crouching or kneeling

Height and weight in manipulating objects

FOUR FEET OR MORE		BETWEEN FOUR FEET AND 18 INCHES		LESS THAN 18 INCHES	
32	0–5 lbs	38	0–5 lbs	44	0–5 lbs
33	0–10 lbs	39	0–10 lbs	45	0–10 lbs
34	0–25 lbs	40	0–25 lbs	46	0–25 lbs
35	0–50 lbs	41	0–50 lbs	47	0–50 lbs
36	50 + lbs	42	50 + lbs	48	50 + lbs
37S	Storage	43S	Divisibility	49S	Storage

Handling objects

50S 51 Manipulating — both

52 Manipulating — either

53S 54 Reaching over 15 inches

55S 56 Reaching above the shoulder

57S 58 Fingering

59S 60 Lifting bulky objects 24 inches or more wide

Moving objects

Carrying:

61	1–5 lbs	64	1–50 lbs	68	Push/pull — one hand	72	5–10 lbs
62	1–10 lbs	65	50 + lbs	69	Push/pull — two hands	73	5–25 lbs
63	1–25 lbs	66	10 lbs beyond 30 feet	70	Push/pull — unequal	74	5–50 lbs
		67	Divisibility	71	Storage	75	50 + lbs

Speech and hearing

[X] 76 Speaking in person
[X] 77 Hearing in person
[ ] 78 Speaking on phone
[ ] 79 Hearing on phone
[ ] 80 Conversation 50% of time with the public
[ ] 81 Hearing — full acuity

Driving and/or machine control placement

[ ] 82 Left-hand control
[X] 83 Right-hand control
[ ] 84 Foot Controls/pedals
[ ] 85 Treading while sitting
[ ] 86 Treading while standing

Infrequent actions

[X] 87 Climbing with legs only
[ ] 88 Climbing with arms and legs
[ ] 89 Twisting the head
[X] 90 Twisting the back

[X] 91 Must crouch
[ ] 92 Must stoop
[X] 93 Must kneel
[ ] 94 Crawling
[ ] 95 Reclining
[ ] 96 Jumping
[ ] 97 Rushing
[ ] 98 Running
[ ] 99 Throwing

Visual Demands Checklist

COMMUNICATION	MEASUREMENT/MANIPULATION	MOBILITY
100 Reading and writing	104 Established method	106 Driving
~~101~~ Working distance	105 Measuring devices	~~107~~ Walking
102 Regular format		~~108~~ Safety hazards
103 Mentoring		

Classification by Strength Requirements Form

DOL strength categories

109 Sedentary	110 Light	~~111~~ Medium	112 Heavy

Duration of walking, standing and sitting

Walking	113	114	115	~~116~~	117	118	119
Standing	120	121	122	123	124	125	126
Sitting	127	128	129	130	131	132	133

Extended or heavy demands

134 Voice	~~136~~ Shoulder	138 Whole body
135 Voice	137 Back and knees	~~139~~ Whole body

Driving

- 140 Assigned
- 141 Regular-assigned
- 142 Short trips
- 143 Major duty
- 144 Central
- 145 May use own vehicle

Physical Barriers Form

- 146 Parking
- 147 Route
- 148 Sign
- 149 12 feet

- 150 Ramp
- 151 48″ wide
- 152 Slope
- 153 Handrail
- 154 Landings

- 155 Entrance
- 156 Door
- 157 32″ wide
- 158 Space
- ~~159~~ Threshold

- 160 Elevator
- 161 Access
- 162 Floor
- 163 Controls
- 164 Braille

- ~~165~~ Floors

- 166 Door
- 167 Handles
- 168 Threshold
- 169 Passage
- ~~170~~ Obstruction

- 171 Rest room
- 172 32″ wide
- 173 Passage
- 174 Vestibule
- 175 Sink
- ~~176~~ Controls

- 177 Stall
- 178 Door
- 179 36″ wide
- 180 Distance
- 181 Grab bar
- 182 Parallel

- 183 Fountain
- 184 Clearance
- 185 Controls

- 186 Telephone
- 187 Public
- ~~188~~ 54″
- ~~189~~ 30″ wide

- 190 Travel

Comments

Job Analyst ____________________

Reviewed by ____________________

Date of job analysis ____________________

Physical Demands Job Analysis

Environmental and Social Conditions Form

Employer job title *Dishwasher*

Instructions — Check any of the conditions listed below that are present in the work area.

Environmental conditions —

CONDITION	DEFINITION
[X] 1 Inside	Protection from weather conditions such as factory, office or long-distance truck driver.
[ ] 2 Outside	No effective protection from the weather such as letter carrier and farmhand.
[X] 3 Extreme heat	Temperature sufficiently high to cause marked bodily discomfort such as working next to a hot stove or furnace or by a hot asphalt spreading machine.
[ ] 4 Extreme cold	Temperature sufficiently low as to cause marked bodily discomfort such as while working in a cold storage room.
[X] 5 Humid or wet	Atmospheric conditions with moisture content sufficiently high to cause marked bodily discomfort such as using a steam garment presser. Also contact with water or other liquids.
[ ] 6 Noise	(1) Sufficient sound to distract workers engaged in mental occupations. This would generally include sounds greater than those made by a single typewriter or office machine. (2) Sound loud enough to potentially cause hearing damage. This would include noise greater than 85 dB which is approximately equal to being in an automobile in heavy city traffic.

No.	Hazard	Description
7	Vibration	Oscillating movement or strain on the body or extremities from repeated motion or shock which may cause harm if endured day after day. Examples include operating a tractor to scoop earth or operating a compressed air rock-drilling machine.
8	Mechanical hazards	Danger to fingers or limbs due to feeding or operating power-driven equipment, either portable or stationary such as a tree chipper or power shearer.
9	Electrical hazards	Danger due to possible electrical shock from electric wires, transformers or uninsulated electrical parts.
10	Hot Material Fire Chemical agents	Danger due to possible burns from any of these causes. (Circle the specific hazard present.)
11	Radiant energy	Exposure to radiant energy such as X-rays, radioactive material, ultraviolet or infrared rays.
12	Poor ventilation	Insufficient or excessive movement of air or exposure to drafts.
13	Moving objects	Exposure to moving equipment and objects in the immediate work area such as automobiles, cranes, heavily laden carts, gurneys and forklifts.
14	Sharp tools	Exposure to tools or material with sharp edges which may involve the risk of injury.
15	Cluttered floors or slippery floors	Walking surface of the work site necessarily strewn with equipment, tools, electrical wiring or other materials while work is being done and which may involve a risk of tripping or falling. Also wet, muddy, greasy, oily or highly polished surfaces which may cause employees to slip or lose footing.
16	Elevated surfaces	Work occurs in places elevated above the ground such as catwalks, scaffolds, ladders and roofs.
17	Poor lighting	Unadjustable lighting conditions which place excessive strain on vision while performing work. Light conditions may be either too dim or too bright causing glare.

[18] (crossed)	Exposure to: fumes odours dust [mist] gases	Circle specific conditions(s) if present.

SOCIAL CONDITIONS

[19] (crossed)	Working with others	Job duties which require communication and/or co-operation with other workers, customers or the public. Includes social skills and ability to tolerate frustration in dealing with others such as sales clerks, clerical supervisors and police.
[20]	Working around others	Job duties which require that the employee work independently but near other employees, or in joint simultaneous effort with co-workers requiring only minimal verbal contact such as carpenter on a construction crew and many assembly line workers.
[21]	Working alone	Work requiring independent occupational effort and virtually no contact with fellow workers or the public such as writer and trapper.
[22]	Supervision	Job duties require direct supervision of other workers.
[23]	Contact with violent or belligerent persons	Job brings worker into contact with individuals who may behave or speak in aggressive, hostile or threatening manner. Also, contact with individuals who have a high probability of such behaviour like police, bouncer and psychiatric aide.
[24]	Working shifts which do not occur between 8 a.m. and 5 p.m.	Working on shifts which do not occur during the 8–5, 9–5 or 7–4 period during the day. Also includes split shifts, rotating shifts and mandatory overtime. Does not include flex-time positions where the employee chooses to work earlier or later.
[25] (crossed)	Protective equipment	Equipment which the workers regularly wear to protect themselves while working. Includes eye protection, canvas or rubber gloves, aprons, safety shoes, hard hats, brightly coloured vests and the like.

Physical Demands Job Analysis

Job Tasks Form .

Employer job title *Dishwasher*

Critical	% of time	Job tasks
✓	50%	1. *Organizes, washes, and resupplies waitresses with dishes and silverware: Pushes button to activate conveyer to bring dirty dishes to sink; Puts bones in bucket, other garbage in disposal, and dishes and utensils into dishwasher basket in sink; Pre-washes dishes with hot water using hand lever operated sprayer. Places basket in dishwasher and adds soap to dishwasher dispenser; Tends semi-automatic dishwasher by pushing 'on' or 'off' button; Removes utensils from dishwasher and stacks dishes and separates silverware for waitresses*

✓	30%	2.	*Washes and resupplies cooks with pots and pans: Fills one sink with hot soapy water and two sinks with hot clear water; Dumps food from pots and pans into disposal and pre-washes them using hand lever operated sprayer; Holds pot or pan in soapy water with one hand and scrubs with brush with other hand. Checks pot or pan for cleanliness and dips in both sinks if clean; Dries pot or pan with dish towel; Changes water when necessary; Places pots and pans on cart and pushes to cooks' area to be stacked.*
✓	20%	3.	*Cleans and maintains premises and equipment: Sweeps and mops floor; Washes dishwasher, all sinks, counters and cart with dishrag and hot soapy water; Carries out trash; Changes light bulbs.*

✓ 10% 4. *Assists cook in stocking walk-in refrigerator: Unloads truck and places boxes, etc., on cart; Pushes cart into referigerator; Stocks items on shelves according to instructions from cook.*

Physical Demands Job Analysis

Task Analysis Form

Job task *Organizes, washes, and resupplies waitresses with dishes and silverware* % of time *50%*

Elements

1. *Organizes, washes and resupplies waitresses with dishes and silverware: Pushes button to activate conveyer to bring dirty dishes to sink; Puts bones in bucket, other garbage in disposal, and dishes and utensils into dishwasher basket in sink; Pre-washes dishes with hot water using hand lever operated sprayer. Places basket in dishwasher and adds soap to dishwasher dispenser; Tends semi-automatic dishwasher by pushing 'on' or 'off' button; Removes utensils from dishwasher and stacks dishes and separates silverware for waitresses*

Tools or work aides used		Machines or equipment used	
	1. *sprayer*		1. *conveyer belt*
	2. *basket for dishes*		2. *sink with disposal; counters*
	3. *soap*		3. *dishwasher*

Most common posture

27S | 26 Walking (or mobility) | ~~28~~ Standing

29 Sitting

30S | 31 Stooping, crouching or kneeling

Height and weight in manipulating objects

FOUR FEET OR MORE		BETWEEN FOUR FEET AND 18 INCHES		LESS THAN 18 INCHES	
32	0–5 lbs	38	0–5 lbs	44	0–5 lbs
33	0–10 lbs	39	0–10 lbs	45	0–10 lbs
34	0–25 lbs	40	0–25 lbs	46	0–25 lbs
35	0–50 lbs	41	0–50 lbs	47	0–50 lbs
36	50 + lbs	42	50 + lbs	48	50 + lbs
37S	Storage	43S	Divisibility	49S	Storage

Handling objects

50S	51	Manipulating — both	53S	54	Reaching over 15 inches	57S	58	Fingering
	52	Manipulating — either	55S	56	Reaching above the shoulder	59S	60	Lifting bulky objects 24 inches or more wide

Moving objects

Carrying:

61	1–5 lbs	64	1–50 lbs	68	Push/pull — one hand	72	5–10 lbs
62	1–10 lbs	65	50 + lbs	69	Push/pull — two hands	73	5–25 lbs
63	1–25 lbs	66	10 lbs beyond 30 feet	70	Push/pull — unequal	74	5–50 lbs
		67	Divisibility	71	Storage	75	50 + lbs

Speech and hearing

- 76 Speaking in person
- 78 Speaking on phone
- 80 Conversation 50% of time with the public
- 77 Hearing in person
- 79 Hearing on phone
- 81 Hearing — full acuity

Driving and control placement

- 82 Left-hand control
- [X] 83 Right-hand control
- 84 Foot controls/pedals
- 85 Treading while sitting
- 86 Treading while standing

Infrequent actions

- 87 Climbing with legs only
- 88 Climbing with arms and legs
- 89 Twisting the head
- [X] 90 Twisting the back
- 91 Must crouch
- 92 Must stoop
- [X] 93 Must kneel
- 94 Crawling
- 95 Reclining
- 96 Jumping
- 97 Rushing
- 98 Running
- 99 Throwing

Notes on physical demands

Worker stands where conveyer belt brings dishes, scrapes and sorts; Pushes basket into dishwasher (during this part the employee stands in one spot; shifting his weight and twisting as needed). He removes the clean dishes, silverware, etc., sorts and stacks. Stacks of plates weigh up to 25 lbs.; Kneeling is needed to store some utensils.

Physical Demands Job Analysis

Task Analysis Form

Job task *Washes and resupplies cooks with pots and pans* % of time *30%*

Elements

2. Washes and resupplies cooks with pots and pans: Fills one sink with hot soapy water and two sinks with hot clear water; Dumps food from pots and pans into disposal and pre-washes them using hand lever operated sprayer; Holds pot or pan in soapy water with one hand and scrubs with brush with other hand. Checks pot or pan for cleanliness and dips in both sinks if clean; Dries pot or pan with dish towel; Changes water when necessary; Places pots and pans on cart and pushes to cooks' area to be stacked

Tools or work aides used		Machines or equipment used
1. brush	*4. towel*	*1. sinks with disposal*
2. soap		*2. counters*
3. sprayer		*3. cart*

Most common posture

[26] Walking (or mobility)
[27S] [~~28~~ X] Standing
[29] Sitting
[30S] [31] Stooping, crouching or kneeling

Height and weight in manipulating objects

FOUR FEET OR MORE		BETWEEN FOUR FEET AND 18 INCHES		LESS THAN 18 INCHES	
32	0–5 lbs	38	0–5 lbs	44	0–5 lbs
33	0–10 lbs	39	0–10 lbs	45	0–10 lbs
34	0–25 lbs	~~40~~ (X)	0–25 lbs	46	0–25 lbs
35	0–50 lbs	41	0–50 lbs	47	0–50 lbs
36	50 + lbs	42	50 + lbs	48	50 + lbs
37S	Storage	43S	Divisibility	49S	Storage

Handling objects

50S ~~51~~ (X) Manipulating — both	53S ~~54~~ (X) Reaching over 15 inches	57S 58 Fingering
52 Manipulating — either	55S 56 Reaching above the shoulder	59S 60 Lifting bulky objects 24 inches or more wide

Moving objects

Carrying:							
61	1–5 lbs	64	1–50 lbs	68	Push/pull — one hand	~~72~~ (X)	5–10 lbs
62	1–10 lbs	65	50 + lbs	~~69~~ (X)	Push/pull — two hands	73	5–25 lbs
~~63~~ (X)	1–25 lbs	66	10 lbs beyond 30 feet	70	Push/pull — unequal	74	5–50 lbs
		67	Divisibility	71	Storage	75	50 + lbs

Speech and hearing

- [76] Speaking in person
- [77] Hearing in person
- [78] Speaking on phone
- [79] Hearing on phone
- [80] Conversation 50% of time with the public
- [81] Hearing — full acuity

Driving and control placement

- [82] Left-hand control
- [83] Right-hand control
- [84] Foot controls/pedals
- [85] Treading while sitting
- [86] Treading while standing

Infrequent actions

- [87] Climbing with legs only
- [88] Climbing with arms and legs
- [89] Twisting the head
- [90] Twisting the back
- [91] Must crouch
- [92] Must stoop
- [93 — crossed out] Must kneel
- [94] Crawling
- [95] Reclining
- [96] Jumping
- [97] Rushing
- [98] Running
- [99] Throwing

Notes on physical demands

Worker stands in front of sink and washes pots and pans weighing up to 20 lbs. each. Pans are dried by hand and stacked for the cook.

Physical Demands Job Analysis

Task Analysis Form

Job task *Cleans and maintains premises and equipment* % of time *20%*

Elements

3. Cleans and maintains premises and equipment: Sweeps and mops floor; Washes dishwasher, all sinks, counters, and cart with dishrag and hot soapy water; Carries out trash; Changes light bulbs

Tools or work aides used		Machines or equipment used
1. push brooms	*4. soap*	*1. mop bucket*
2. mops	*5. dustpans*	*2. ladder*
3. rags		*3. trash cans*

Most common posture

[27S] [~~26~~] Walking (or mobility) [28] Standing [29] Sitting [30S] [31] Stooping, crouching or kneeling

Height and weight in manipulating objects

FOUR FEET OR MORE		BETWEEN FOUR FEET AND 18 INCHES		LESS THAN 18 INCHES	
~~32~~	0–5 lbs	38	0–5 lbs	~~44~~	0–5 lbs
33	0–10 lbs	39	0–10 lbs	45	0–10 lbs
34	0–25 lbs	40	0–25 lbs	46	0–25 lbs
35	0–50 lbs	~~41~~	0–50 lbs	47	0–50 lbs
36	50 + lbs	42	50 + lbs	48	50 + lbs
37S	Storage	43S	Divisibility	49S	Storage

Handling objects

50S	~~51~~	Manipulating — both	53S	54	Reaching over 15 inches	57S	58	Fingering
	52	Manipulating — either	55S	~~56~~	Reaching above the shoulder	59S	~~60~~	Lifting bulky objects 24 inches or more wide

Moving objects

Carrying:

61	1–5 lbs	~~64~~	1–50 lbs	68	Push/pull — one hand	72	5–10 lbs
62	1–10 lbs	65	50 + lbs	69	Push/pull — two hands	73	5–25 lbs
63	1–25 lbs	66	10 lbs beyond 30 feet	~~70~~	Push/pull — unequal	~~74~~	5–50 lbs
		67	Divisibility	71	Storage	75	50 + lbs

Speech and hearing

- 76 Speaking in person
- 77 Hearing in person
- 78 Speaking on phone
- 79 Hearing on phone
- 80 Conversation 50% of time with the public
- 81 Hearing — full acuity

Driving and control placement

- 82 Left-hand control
- 83 Right-hand control
- 84 Foot controls/pedals
- 85 Treading while sitting
- 86 Treading while standing

Infrequent actions

- ☒ 87 Climbing with legs only
- 88 Climbing with arms and legs
- 89 Twisting the head
- ☒ 90 Twisting the back
- 91 Must crouch
- ☒ 92 Must stoop
- 93 Must kneel
- 94 Crawling
- 95 Reclining
- 96 Jumping
- 97 Rushing
- 98 Running
- 99 Throwing

Notes on physical demands

After floors are swept, worker stoops to use dustpan; Mop and mop bucket are pushed to work area; Mopping requires twisting of lower back; Washing fixtures and dishwasher requires no stooping and reaching: Trash cans weigh up to 50 lbs.; Changing light bulbs requires the use of a ladder.

Physical Demands Job Analysis

Task Analysis Form

Job task *Assists cook in stocking walk-in refrigerator* % of time *10%*

Elements

4. Assists cook in stocking walk-in refrigerator: Unloads truck and places boxes, etc., on cart; Pushes cart into refrigerator; Stocks items on shelves according to instructions from cook

Tools or work aides used

Machines or equipment used *1. walk-in refrigerator*
2. cart

Most common posture

[27S] [~~26~~] Walking (or mobility) [28] Standing

[29] Sitting

[30S] [31] Stooping, crouching or kneeling

<u>Height and weight in manipulating objects</u>

FOUR FEET OR MORE		BETWEEN FOUR FEET AND 18 INCHES		LESS THAN 18 INCHES	
32	0–5 lbs	38	0–5 lbs	44	0–5 lbs
33	0–10 lbs	39	0–10 lbs	45	0–10 lbs
34 (crossed)	0–25 lbs	40 (crossed)	0–25 lbs	46 (crossed)	0–25 lbs
35	0–50 lbs	41	0–50 lbs	47	0–50 lbs
36	50 + lbs	42	50 + lbs	48	50 + lbs
37S	Storage	43S	Divisibility	49S	Storage

<u>Handling objects</u>

50S	51 (crossed)	Manipulating — both	53S	54 (crossed)	Reaching over 15 inches	57S	58	Fingering
	52	Manipulating — either	55S	56 (crossed)	Reaching above the shoulder	59S	60 (crossed)	Lifting bulky objects 24 inches or more wide

<u>Moving objects</u>

Carrying:

61	1–5 lbs	64	1–50 lbs	68	Push/pull — one hand	72	5–10 lbs
62	1–10 lbs	65	50 + lbs	69 (crossed)	Push/pull — two hands	73 (crossed)	5–25 lbs
63 (crossed)	1–25 lbs	66	10 lbs beyond 30 feet	70	Push/pull — unequal	74	5–50 lbs
		67	Divisibility	71	Storage	75	50 + lbs

Speech and hearing

- [X] 76 Speaking in person
- [ ] 77 [X] Hearing in person
- 78 Speaking on phone
- 79 Hearing on phone
- 80 Conversation 50% of time with the public
- 81 Hearing — full acuity

Driving and control placement

- 82 Left-hand control
- 83 Right-hand control
- 84 Foot controls/pedals
- 85 Treading while sitting
- 86 Treading while standing

Infrequent actions

- 87 Climbing with legs only
- 88 Climbing with arms and legs
- 89 Twisting the head
- 90 Twisting the back
- [X] 91 Must crouch
- 92 Must stoop
- 93 Must kneel
- 94 Crawling
- 95 Reclining
- 96 Jumping
- 97 Rushing
- 98 Running
- 99 Throwing

Notes on physical demands

The worker loads the cart as instructed by cook; Pushes cart into refrigerator and unloads; Boxes weigh 25 lbs., and lighter items are bulky; Reaching and crouching are necesssary to reach all storage areas.

Physical Demands Job Analysis

Visual Demands Checklist

Employer job title Dishwasher

To use this form:

(A) Analyse the physical demands of the duties for the position under consideration.
(B) Complete the Environmental Conditions Form.
(C) Read the manual for complete definitions of the job characteristics listed below.
(D) Make a decision about whether any of the job characteristics listed below seem to be true of the position. Take into account the main and critical job duties, the work environment and any other details from the interview.
(E) The presence of one or more of these characteristics in the work may mean that accommodation regarding the blind might be considered in more detail.

Communication

[100] Essential and extensive reading and writing is generally performed at a few locations.

[101 ☒] Visual working distances are not fixed.

[102] Most of the material read is printed or typed and the material is generally within a regular or standardized format.

[103] Mentoring, particularly in a verbal manner, is the dominant activity.

Measurement/manipulation

[104] Manipulation of objects occurs within a well defined, established method or pattern.

[105] Measuring devices could be converted to tactile, auditory or enlarged scales rather than small print.

Mobility

[106] Driving is limited (or unnecessary) and occasions of use can be controlled by the worker or the supervisor.

[107] The work is not characterized by frequent required walking to a variety of non-routine locations outside the work unit where speed of mobility is essential.

[108] Immediate work area does not contain major safety hazards from moving objects which could result in serious injury.

Physical Demands Job Analysis

Classification by Strength Requirements Form

Employer job title *Dishwasher*

Degrees of strength

[109] SEDENTARY WORK: Lifting 10 lbs maximum and occasionally lifting and/or carrying such articles as ledgers or hand tools. Although a sedentary job is defined as one which involves sitting, a certain amount of walking and standing may be occasionally necessary in performing job duties. Example: worker sits at a desk most of day, takes dictation and transcribes it on a typewriter, and occasionally walks to various departments. Example: worker sits at a drawing board and walks occasionally, and carries paper, instruments and books.

[110] LIGHT WORK: Lifting 20 lbs maximum with frequent lifting and/or carrying objects weighing up to 10 lbs. Even though weight lifted is negligible, a job will be in this catetory when, (1) it requires walking or standing to a significant degree, or (2) when it requires sitting most of the time, but entails pushing or pulling of arm and/or leg controls. Example: worker stands and walks behind a counter of a variety store all of working day wrapping and bagging articles for customers. Example: worker walks and stands constantly while arranging records in file cabinets and sits occasionally to sort paper. Example: worker sits most of day and operates an industrial sewing machine.

[X 111] MEDIUM WORK: Lifting 50 lbs maximum with frequent lifting and/or carrying of objects weighing up to 25 lbs. Example: worker assists in lifting patients, pushing litters and pulling sheets from beds. Example: as the heaviest of several job duties, worker lifts, pushes and pulls, to jack up an automobile and remove tyre from wheel and remounts tyre on wheel.

[112] HEAVY WORK: Lifting 100 lbs maximum with frequent lifting and/or carrying of objects weighing up to 50 lbs. Example: worker pushes handtruck up and down aisles of warehouse to fill orders, stooping and lifting cartons or items with an average weight of 65 lbs and placing them on the truck. Example: worker digs a trench by hand to specified depth and width using a shovel.

Duration of walking, standing and sitting

TIME	DESCRIPTION	WALKING	STANDING	SITTING
$\frac{1}{2}$ Hour or less	brief, intermittent occasions	113	120	127
$\frac{1}{2}$ Hour	regular, but minimal	114	121	128
1 Hour	regular, but minor	115	122	129
2 Hours	less than a third of day	~~116~~	123	130
3 Hours	less than half the day	117	124	131
4–6 Hours	more than half to a majority of the working day	118	125	132
6 or more hours	all of the working day with occasional breaks	119	~~126~~	133

Extended or heavy physical demands

134	Voice — extended	Necessary and extended use of the voice each day as in the position of telephone operator or interviewer.
135	Voice — loud	Duties require using a loud voice.
~~136~~	Shoulders	Prolonged unsupported horizontal extension of the arms as in a stenographer position or frequent reaching above the shoulders as in a stock clerk position.

137	Back and knees	Extended stooping, crouching or kneeling. Also stooping, crouching or kneeling occurring on a regular, repeated basis within a limited time span during the workday which constitutes a heavy physical demand.
138	Whole body — extension	Frequent or extended crawlings, reclining, jumping or climbing with arms and legs.
~~139~~ (crossed)	Whole body — twisting	Duties regularly involve twisting the spine or extending the body while lifting weight in excess of 25 lbs which is located four feet or more off the ground or under 18 inches. Also includes pushing or pulling with 25 lbs of force while twisting or extending.

Need for driving

140 On rare or infrequent occasions, the supervisor may assign a short trip to any one of several staff members.

141 On regular occasions, the supervisor may choose one among several staff members to drive on a short trip.

142 The duties of the position require regular short trips each month.

143 The position requires driving as one major job duty. Driving should be equal to at least one day in each week.

144 Driving is the principal and central job duty of the position.

145 Worker may use own vehicle so that arrangement of controls can be specified by him/her.

Physical Demands Job Analysis

Physical Barriers Form

Employer job title *Dishwasher*

Building *Howard Johnson's Restaurant*

Job location within building *Kitchen*

DARKENING A BOX INDICATES A NEGATIVE RESPONSE TO THE QUESTION

[146] Is PARKING provided which is adjacent or convenient to building?

- [147] Is route from parking to building barrier free?
- [148] Is a space indicated for handicapped?
- [149] Is a space at least 12′ wide?

[150] Is RAMP available for access to building?

- [151] Is walk/ramp at least 48″ wide?
- [152] Is slope no more than 1″ rise to 12″ in length?
- [153] Does ramp have at least one handrail 32″ above surface?
- [154] Does ramp have landings at top and bottom for 5′ in direction of travel?

155 Is at least one accessible BUILDING ENTRANCE a main entrance?

- 156 Does door require less than 8 lbs of force to open?
- 157 Is there a minimum 32″ clear opening?
- 158 In direction of door swing, is there an area at least 5′ by 5′?
- ~~159~~ Is threshold flat or with maximum ½″ slope?

160 Are ELEVATORS directly accessible from a usable entrance?

- 161 Is office/work site directly accessible from usable main entrance or from elevator?
- 162 Does elevator car automatically come within ½″ of building floor?
- 163 Are the highest controls within 54″ of floor?
- 164 Do elevator panels have braille and raised arabic symbols to right of buttons?

~~165~~ Are FLOORS of a smooth, non-slip material or covered with low-nap carpet?

166 Do DOORS within office/work site have a 32″ clear opening?

- 167 Are doors opened by lever handles rather than door knobs?
- 168 Are thresholds sloped and less than ½″ high?
- 169 Are passageways at least 48″ wide and free of tight turns?
- ~~170~~ Can existing obstructions within passageways or office be easily moved?

- [171] Is REST ROOM convenient and accessible to work site?
  - [172] Do entry doors have 32″ clear opening?
  - [173] Is passageway from rest room entrance to stall 43″ wide at all points?
  - [174] If there is a vestibule between two doors, is it at least $5\frac{1}{2}$′ long?
  - [175] Is there at least 29″ of vertical clearance under sink?
  - [176] (crossed out) Are the sink controls the single lever type?
- [177] Is there at least one rest room STALL which has 30″ clear entrance?
  - [178] Does stall door open out?
  - [179] Is stall at least 36″ wide?
  - [180] Is the distance from the front of the stool to the closed door at least 48″?
  - [181] Is there a grab bar on each side of the stall?
  - [182] Are the grab bars parallel to and 33″ from floor?
- [183] Is one WATER FOUNTAIN no more than 36″ from floor, or has a cup dispenser been provided?
  - [184] Is there 27″ knee clearance beneath fountain?
  - [185] Are spout and controls up-front?
- [186] Is TELEPHONE available and accessible for use?
  - [187] If not, is a public phone in an accessible area?
  - [188] (crossed out) Is one phone mounted so that the highest operable parts are no more than 54″ above the floor?
  - [189] (crossed out) Does the booth entrance provide 30″ wide clearance?
- [190] Job requires TRAVEL to non-company sites which may not be accessible.

Appendix B
Rating schemata for WCAM/WPAM analyses

Appendix B comprises selected samples of page content from the WCAM/WPAM manual presented by Nylander and Carmean (1984). These samples are selected to illustrate the layout of different aspects of the Manual and are merely representative. Any person wishing to use the system should consult the Manual. The questions in Part 1 apply to workers in the United States. They are not mandatory and need not be answered if the employee is unwilling.

Part I
Biographical Data

1–5. To be defined by user
6. Sex: M F
7. What do you consider your ethnicity?
   1. White
   2. Black
   3. Hispanic
   4. Asian or Pacific Islander
   5. American Indian or Alaskan Native
8. Which best describes the education you have completed?
   1. Less than High School
   2. High School Diploma or GED
   3. Less than two years of college
   4. AA degree or certificate
   5. Three or four of college — no degree
   6. Bachelor's degree or higher
9. How long have you been in your position?
   1. Less than 3 months
   2. 3–6 months
   3. 7–12 months
   4. More than 1 year but less than 2 years
   5. 2 years but less than 4 years
   6. 6 years or more
10. If you supervise, how many people do you supervise?
11. Do you consider yourself to have one of the following physical conditions?
   1. Impairment of sight, hearing or speech
   2. Impairment of physical ability because of amputation, loss of function or loss of coordination
   3. Any health impairment which requires special education or related services

Part II
Working Conditions

(Note: the following are representative items only. They do not include the complete list.)

1. *Inside*

Working under a roof and with all sides protected from the weather (exclude motor vehicles from consideration).

(0) No effect — Is not a condition of the Study Job
(1) Little — 10% or less of working time spent inside a building
(3) Moderate — 10% to 90% of working time spent inside a building
(4) Great — 90% to 100% of working time spent inside a building

2. *Outside*

Working outside exposed to the weather — heat, cold, humidity, dryness, wetness and dust (due to climate rather than other sources).

(0) No effect — Is not a condition of the Study Job
(1) Little — 10% or less of working time spent inside a building
(2) Moderate — 10% to 90% of working time spent outside a building
(3) Great — 90% to 100% of working time spent outside a building

15. *Noise*

Working condition in which sound is produced as part of the work process or is a part of the job.

(0) No effect — Is not a condition of the Study Job
(1) Little — Low sound or occasional fairly loud sounds
(2) Moderate — Steady and fairly loud noises
(3) Great — Intermittent or continued loud and insistent noise

21. *Toxic Conditions*

Exposure to toxins: dusts (other than silica and asbestos), fumes, liquids, gases (aldehydes, other than gases resulting from plastics fires; or carbon monoxide, the effects of which may be multiplied by smoking or proximity to open flame) which cause general or localized disabling conditions.

(0) No effect — Is not a condition of the Study Job
(1) Little — Exposure limited, or substances low in concentration
(3) Moderate — Exposure frequent, or substances fairly high in concentration
(4) Great — Sustained exposure to toxic substances whose effects may be cumulative, or substances very high in concentration or which may require wearing of self-contained breathing apparatus

33. *Role Ambiguity*

Lack of clarity about what others expect of you on the job.

(0) No effect — Is not a condition of the Study Job
(1) Little — Rarely is it not clear what others expect of you on the job
(2) Moderate — Occasionally what is expected is not clear
(3) Great — What is expected is usually not clear

34. *Irregular or extended work hours*
Working under conditions that cause fluctuating work hours.

(0) No effect — Is not a condition of the Study Job
(1) Little — Relatively slight chance that work hours will change or you will be required to work beyond normal hours
(2) Moderate — Occasionally required to change work hours or work beyond normal quitting time
(3) Great — Frequently required to change (rotate) work hours (shift assignment) or to extend work hours beyond normal quitting time

Part III
Physical Abilities

1. *Static Strength*
This is the ability to use muscle force to lift, push, pull or carry objects. It is the maximum force that one can exert for a brief period of time. This ability can involve the hand, arm, back, shoulder or leg.

How Static Strength is Different from Other Abilities

Use muscle to exert force against objects	vs.	Dynamic strength and trunk strength: Use muscle power repeatedly to hold up or move the body's own weight
Use *continuous* muscle force, without stopping, up to the amount needed to lift, push, pull or carry an object	vs.	Explosive strength: gather energy to move one's own body, to propel some object with *short bursts* of muscle force
Does *not* involve the use of muscle force over a long time	vs.	Stamina: *does* involve physical exertion over a long time

Requires use of all the muscle force possible	7—	
	6—	Lift bags of cement into truck
	5—	
	4—	
		Pull sack of mulch across a yard
	3—	
	2—	
Requires use of a little muscle force to lift, carry, push or pull an object	1—	Lift a package of bond paper

6. *Effort*

This is the degree of physical exertion experience in performing either a single task or a series of tasks.

Requires extensive physical exertion	7—	
		Operate a jackhammer
	6—	
	5—	
	4—	
		Perform light welding
	3—	
	2—	
Requires little physical exertion	1—	Sit at desk using a hand calculator

14. *Manual dexterity*

This is the ability to make skilful, co-ordinated movements of one hand, a hand together with its arm or two hands. These movements are used to grasp, place, move or assemble objects like hand tools or blocks. This ability involves the degree to which these arm–hand movements can be carried out quickly. It does not involve moving machine or equipment controls like levers.

How Manual Dexterity is Different from Other Abilities

Involves the *co-ordination* of the arm and hand	vs.	Arm–hand steadiness: involves *steadiness* of the arm and hand
Skilled movements involve mainly the *hands*		Finger dexterity: skilled movements involving mainly the *fingers*

Requires very fast, skilful use of one hand, a hand and arm, or two hands to grasp, place, move or assemble objects	7—	Perform open heart surgery
	6—	
	5—	Repair upholstery with hand tools
	4—	Package oranges in crates as rapidly as possible Prune shrubs with shears
	3—	
	2—	
		Turn handle on tank to start flow
Requires some speed, skill and co-ordination to grasp, place, move or assemble with one hand, hand and arm, two hands	1—	

16. *Near Vision*

The capacity to see close environmental surroundings.

Requires long periods of directed near visual activity 7—

6—

Read the fine print of legal journals

5—

Cut and mount color film transparencies

4—

3—

2—

Plug in a TV set

Requires occasional near activity 1—

Appendix C
Physiological and biomechanical techniques for work capacity measurement

Cardiopulmonary function

Muscular exercise is an energy-consuming activity. The basic energy source of the body is glucose which is stored in the muscles in the form of glycogen. For immediate use, as in the initiation of exercise, the glycogen is broken down in the body by the action of chemical catalysts known as enzymes through several stages to lactic acid, thereby releasing stored energy in the process. Since this process does not require oxygen it is called anaerobic. As more energy is required the process is continued with the use of oxygen, and still more energy is released by way of aerobic oxidation. In continued mild to moderate exercise the oxygen supply is sufficient to meet the demands of oxidation. Should more oxygen be required an oxygen debt is built up which is recovered by panting and overbreathing after the exercise is completed.

It will be clear that one of the main essentials in the performance of exercise is the supply of oxygen. The maximal oxygen uptake or the maximal aerobic power is defined as the highest oxygen uptake the person can attain during physical work breathing air at sea level. Mean values for maximal oxygen uptake have been measured for numerous groups including the 350 male and female subjects from 4 to 65 years of age whose data are shown in Figure C.1. It will be noted that values vary with age, sex and level of physical training. The information provided by the assessment of maximal oxygen uptake is a measure of the maximal energy output by way of aerobic processes, and also the functional capacity of the circulation, since there is a high correlation between the maximal cardiac output and the maximal aerobic power (Astrand and Rodahl, 1970).

Three methods are in use for standardizing the workload for the measurement of maximal oxygen uptake. These are firstly, running at a controlled rate under conditions of variable slope on a treadmill; secondly, cycling, either by pedal or by hand crank on a bicycle ergometer, which is a type of stationary bicycle on which the load on the back wheel can be varied and from which the extent of the load can be read or calculated; and thirdly, stepping up and down one or two steps through a controlled distance in a step-test. In general, the work should involve large muscle groups and the measurement of the oxygen uptake should be started when the work has lasted a few minutes.

Most exercise physiologists prefer the use of a bicycle ergometer or treadmill because the output is more controllable and the subject has less motion, but the step-test requires simple equipment and may be more familiar to the subject.

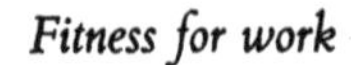

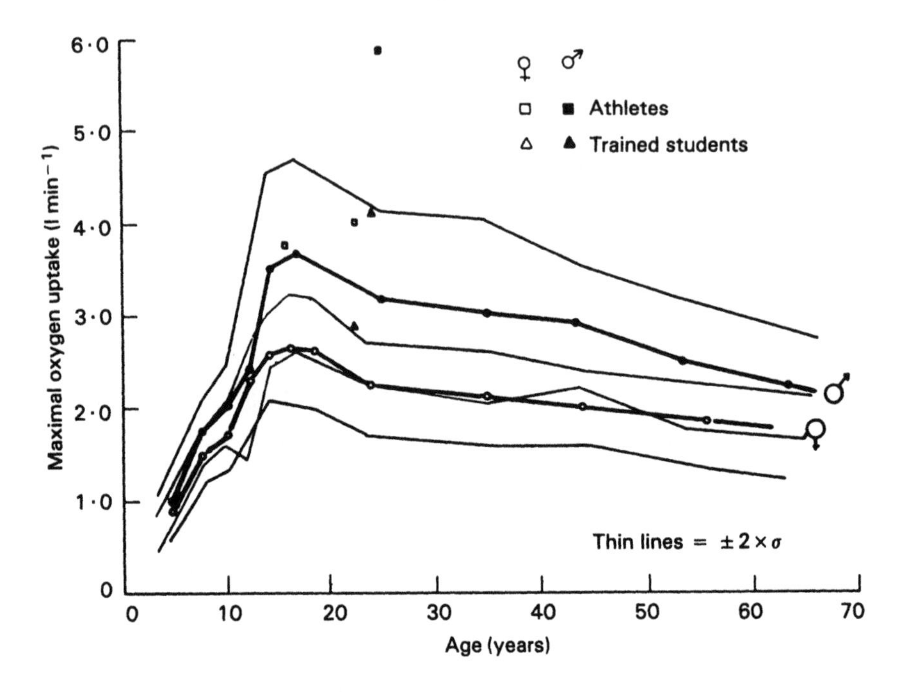

Figure C.1. Mean values for maximal oxygen uptake by age — 358 subjects (Astrand and Christensen, in Astrand and Rodahl, 1970)

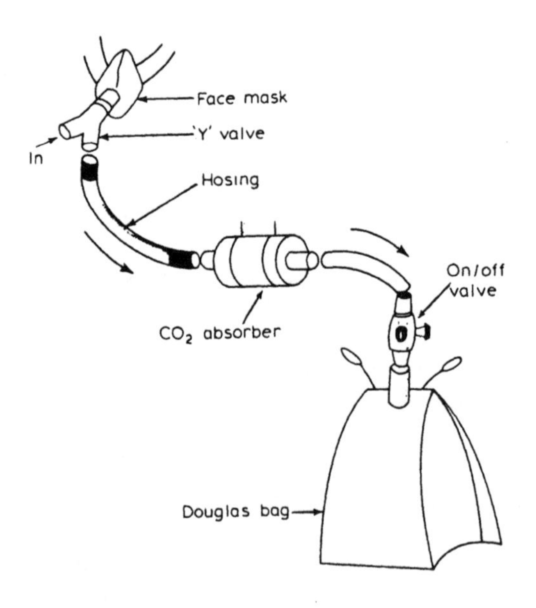

Figure C.2. Schematic model of system for collecting expired air (Ayoub, 1983)

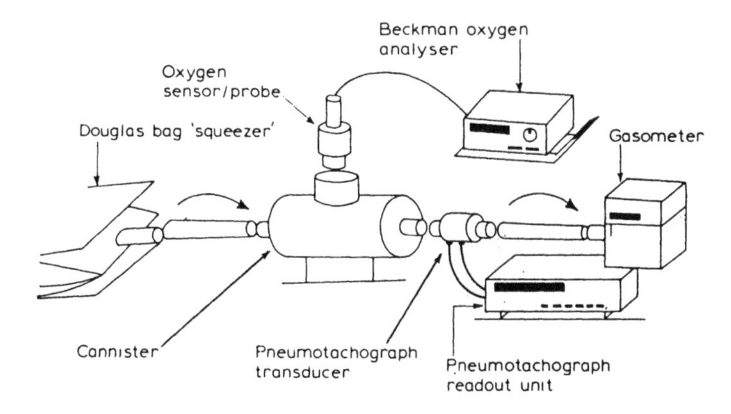

Figure C.3. Schematic model of system for analysing expired air (Ayoub, 1983)

Contemporary step-tests are derivatives of the Harvard Step-Test which was developed during World War II as a fitness screening test. The height of the bench was 50 cm and the stepping frequency 30 steps min^{-1}, so selected that only about one-third of subjects would be expected to complete the test over a period of 5 min. As a screening test it tended to predict those who could perform well in activities that demanded a high aerobic power, but it failed to take into account other factors such as anaerobic power, technique and motivation (Astrand and Rodahl, 1970). As a means of producing a relatively controlled exercise demand, however, it is useful in the prediction of maximal oxygen uptake.

Determination of oxygen uptake is commonly undertaken by gas analysis of expired air. A system for this purpose is illustrated in schematic form in Figures C.2 and C.3. Essentially, expired air is collected via a face mask fitted with a Y-valve leading to a flexible hose and then to a large airtight Douglas bag with a canister for the absorption of carbon dioxide. The bag can be carried on the back in a moving subject or suspended from a harness. When at least 50 l of air have been collected over seconds to minutes, as appropriate, the system is disconnected. The air from the Douglas bag is then expelled for measurement of volume and gas analysis. The gas analysis can be done by the traditional Haldane or Micro-Scholander wet chemical methods, or by way of a proprietary electronic gas analyser such as the Beckman Oxygen Analyser (Model 0260). Volumes can be measured by pneumotachograph (e.g. Hewlett-Packard) and gasometer (e.g. Max Planck). Gas collection is commonly accompanied by the measurement of heart rate and rhythm using an electrocardiograph and/or cardiotachometer.

It will be apparent that the use of such a system is complex and requires the services of skilled technicians or assistants. While it is certainly appropriate in an exercise laboratory or a scientific field test, and perhaps even for a large industrial corporation with suitable facilities it would not be suitable for use in an ordinary occupational health setting.

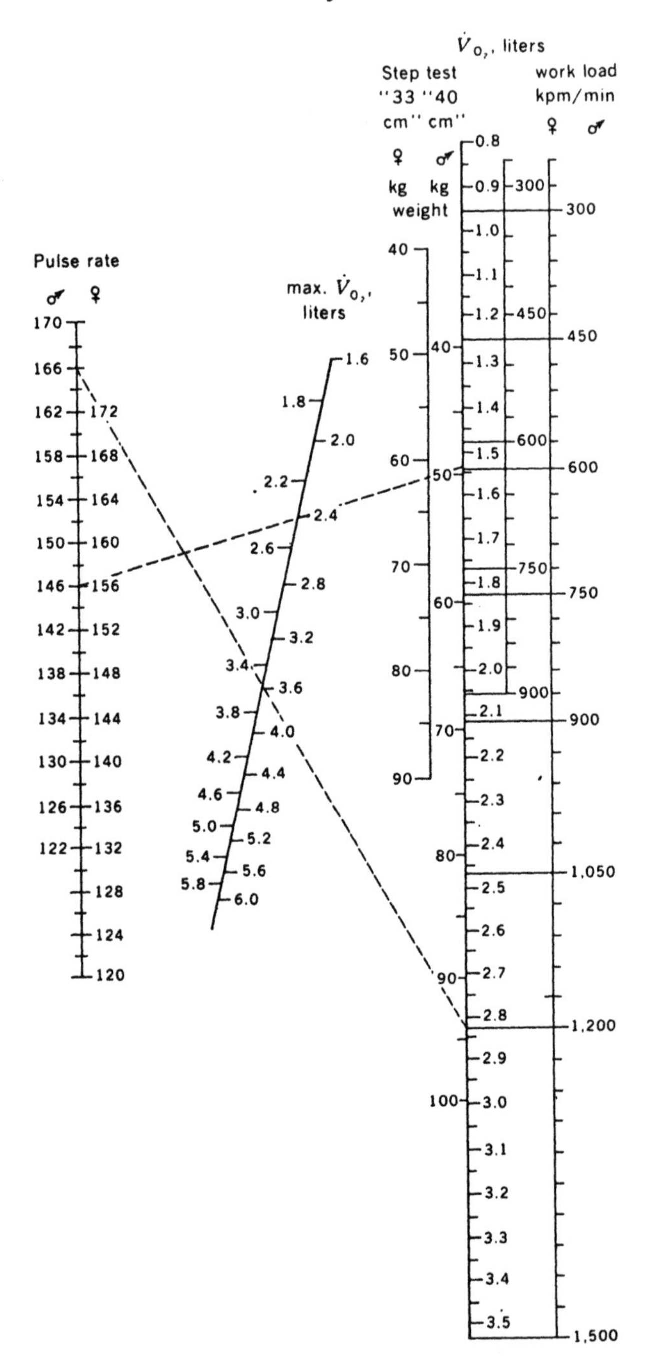

Figure C.4. A nomogram for determining pulse rate, predicted oxygen uptake, step test conditions, work, subject weight, and actual oxygen uptake (Astrand and Rodahl, 1954)

It is fortunate, however, that a direct correlation can be demonstrated between heart rate in exercise and maximal oxygen uptake, such that within some reasonable limits of error the heart rate can be used for predictive purposes, while also being used for assessing circulatory capacity.

The simplest and most extensively applied way of testing is to determine the heart rate during or after exercise on a bicycle ergometer, treadmill or step-test. Normally the recording should take place during work, but in the individual subject there is a high correlation between pulse rate taken while working and pulse rate taken 1–1·5 min after working. It has been demonstrated on numerous occasions that there is a linear increase in heart rate with increasing oxygen uptake or workload. On the basis of one such study (Astrand and Rhyming, 1954), a nomogram was constructed as shown in Figure C.4 to account for pulse rate, predicted maximum oxygen uptake, step-test conditions, workload, subject weight and actual oxygen uptake.

From Figure C.4 it will be seen that in tests without direct measurement of oxygen uptake, the latter can be estimated by connecting a line from the body weight scale (step-test) or the workload scale (ergometer test) to the oxygen uptake scale. From there a line connected to the appropriate point on the pulse rate scale will cross the maximal oxygen uptake scale at a point representing predicted maximal oxygen uptake. Since maximum heart rate reduces with age a correction for age has to be made by multiplying the nomogram values by a factor obtained from Table C.1. The standard error of the method is considered to be between 10 and 15%.

Such tests should be done under 'steady-state' conditions, that is, during continuous steady work lasting at least 5–6 min. The load should be chosen so that the maximum heart rate reaches at least 140 beats min^{-1} in the case of subjects below 50 years of age, and 120 beats min^{-1} in persons above 50 years of age. For trained fit male subjects, this load would be 600–900 kiloponds min^{-1} on a bicycle ergometer, and for trained females 450–600 kiloponds min^{-1}. For older untrained persons a load of about 300 kiloponds min^{-1} would be suitable. The subject should refrain from energetic physical activity for 2 h preceding the test, which should not be performed less than 1 h after a light meal or 2 h after a heavy meal. Normally, unless under special conditions with a physician present, the exercise should be stopped when the heart

Table C.1. Factor to be used for predicted maximal oxygen uptake, when subject is 30–35 years or over or when subject's heart rate is known (Astrand and Rodahl, 1970).

Age	Factor	Maximum heart rate	Factor
15	1·10	210	1·12
25	1·00	200	1·00
35	0·87	190	0·93
40	0·83	180	0·83
45	0·78	170	0·75
50	0·75	160	0·69
55	0·71	150	0·64
60	0·68		
65	0·65		

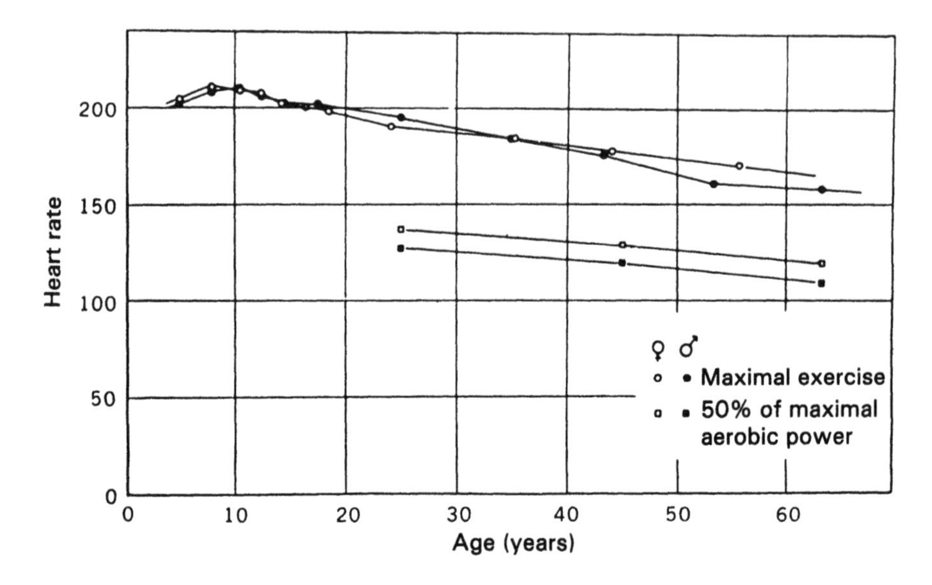

Figure C.5. Heart rate during manual exercise (upper curve), during prolonged work representing 50% of an individual's maximal oxygen uptake (lower curve) for subjects in Figure C.1 (Astrand and Christensen in *Astrand and Rodahl, 1970*)

rate reaches 150 beats min^{-1}. A working pulse rate of 150 beats min^{-1} represents an almost maximal effort for an untrained person with a maximal pulse rate of 160 beats min^{-1}, although it might represent a relatively light load for a trained person with a maximal rate of 200 beats min^{-1}.

Providing the work is not too heavy, respiration and circulation normally increase during the first 2–3 min of work and then attain a steady state. Pulse rate tends to reach steady state after 4–5 min of work. The rate should be measured by palpation of the carotid or radial pulse every minute, the mean value of the fifth and sixth minute being designated as the working pulse rate.

Heart rate during maximal exercise would be expected to fall within the values shown in Figure C.5, which shows the maximal heart rates recorded on 350 mixed subjects during maximal exercise. Maximal oxygen uptake would be expected to be in the range demonstrated in Figure C.1, from the same group of subjects.

The concept of heart rate utilization is used by Ayoub (1982, 1983) in determining the ability of job applicants to meet work output requirements. He has developed a number of charts to assist in determining that work output. As input to the chart shown in Table C.2 (a) he takes the difference between the heart rate following completion of the lower load step-test and that following the higher load and draws a line horizontally to meet a descending line from the subject's weight. The resulting number defines the expected increment in heart rate that would occur from the load defined on the chart. This number is added to the lower level heart rate to define the actual predicted heart rate at that load. The predicted heart rate as so defined is used as input to Table C.2 (b). A line extrapolated from that input to a descending line from

Table C.2(a). Example of chart for predicting heart rate increment. Increment for person weighing 179 lbs, with 12 beats per minute (bpm), between high and low levels of step test is 43 bpm (Ayoub, 1983).

Difference between step test heart rates	Predicted job heart rate Weight (lbs)																				Job workload 500 weight 100–204			
	100–3	104–7	108–11	112–5	116–20	121–4	125–9	130–4	135–9	140–3	144–7	148–52	153–7	158–61	162–5	166–9	170–3	174–7	178–82	183–7	188–92	193–6	197–200	201–4
5	28	27	26	25	24	25	24	23	22	22	21	22	22	27	26	25	25	24	24	23	22	24	24	24
6	32	31	30	28	27	28	27	26	25	24	24	25	24	30	29	29	28	27	26	26	25	27	27	26
7	36	34	33	32	30	31	30	28	27	27	26	27	26	33	33	32	31	30	29	28	27	30	30	29
8	39	38	36	35	33	34	33	31	30	29	28	30	29	37	36	35	34	33	32	31	30	33	32	32
9	43	41	39	38	36	37	35	34	32	31	30	32	31	40	39	38	37	36	35	33	32	36	35	34
10	47	45	43	41	39	40	38	36	35	34	33	35	33	44	42	41	40	39	37	36	35	39	38	37
11	51	48	46	44	41	43	41	39	37	36	35	37	36	47	45	44	43	41	40	39	37	42	41	40
12	54	52	49	47	44	46	44	42	40	36	37	39	38	50	49	47	46	44	43	41	40	45	43	42
13	58	55	52	50	47	49	47	44	42	41	39	42	40	54	52	50	49	47	45	44	42	47	46	45
14	62	58	56	53	50	52	49	47	45	43	42	44	43	57	55	53	52	50	48	46	45	50	49	48
15	65	62	59	56	53	55	52	50	47	45	44	47	45	60	58	56	55	53	51	49	47	53	52	51
16	69	65	62	59	56	58	55	52	50	48	46	49	47	64	62	60	58	56	54	51	50	56	55	53
17	73	69	65	62	59	61	58	55	52	50	48	52	50	67	65	63	61	59	56	54	52	59	57	56
18	76	72	69	65	61	64	61	58	55	53	51	54	52	70	68	66	64	61	59	57	54	62	60	59
19	80	76	72	68	64	67	63	60	57	55	53	57	54	74	71	69	67	64	62	59	57	65	63	61
20	84	79	75	71	67	70	66	63	60	57	55	59	57	77	74	72	69	67	64	62	59	68	66	64
21	87	83	78	75	70	73	69	65	62	60	57	62	59	80	78	75	72	70	67	64	62	70	69	67
22	91	86	82	78	73	76	72	68	65	62	60	64	61	84	81	78	75	73	70	67	64	73	71	69
23	95	90	85	81	76	79	75	71	67	64	62	66	64	87	84	81	78	76	73	70	67	76	74	72
24	98	93	88	84	79	82	78	73	70	67	64	69	66	91	87	84	81	79	75	72	69	79	77	75
25	102	97	92	87	81	85	80	76	72	69	66	71	68	94	91	87	84	81	78	75	72	82	80	78
26	106	100	95	90	84	88	83	79	75	71	69	74	71	97	94	90	87	84	81	77	74	85	82	80
27	109	104	98	93	87	91	86	81	77	74	71	76	73	101	97	94	90	87	84	80	77	88	85	83
28	113	107	101	96	90	94	89	84	80	76	73	79	75	104	100	97	93	90	86	83	79	91	88	86

Table C.2(b). Example of chart for assessing heart rate utilization by age and predicted job rate (Ayoub, 1983).

	Heart rate utilisation as a function of age and workload Age (years)															
Predicted job heart rate	19–21	22–24	25–27	28–30	31–33	34–36	37–39	40–42	43–45	46–48	49–51	52–54	55–57	58–60	61–63	64–66
100–101	51	52	52	53	53	54	54	55	55	56	57	57	58	59	59	60
102–103	52	53	53	54	54	55	55	56	57	57	58	58	59	60	60	61
104–105	53	54	54	55	55	56	56	57	58	58	59	60	60	61	62	62
106–107	54	55	55	56	56	57	57	58	59	59	60	61	61	62	63	63
108–109	55	56	56	57	57	58	59	59	60	60	61	62	62	63	64	65
110–111	56	57	57	58	58	59	60	60	61	62	62	63	64	64	65	66
112–113	57	58	58	59	59	60	61	61	62	63	63	64	65	66	66	67
114–115	58	59	59	60	61	61	62	62	63	64	65	65	66	67	67	68
116–117	59	60	60	61	62	62	63	64	64	65	66	66	67	68	69	69
118–119	60	61	61	62	63	63	64	65	65	66	67	68	68	69	70	71
120–121	61	62	62	63	64	64	65	66	66	67	68	69	69	70	71	72
122–123	62	63	63	64	65	65	66	67	68	68	69	70	71	71	72	73
124–125	63	64	64	65	66	66	67	68	69	69	70	71	72	73	73	74
126–127	64	65	66	66	67	68	68	69	70	70	71	72	73	74	75	75
128–129	65	66	67	67	68	69	69	70	71	72	72	73	74	75	76	77
130–131	66	67	68	68	69	70	70	71	72	73	74	74	75	76	77	78
132–133	67	68	69	69	70	71	72	72	73	74	75	75	76	77	78	79
134–135	68	69	70	70	71	72	73	73	74	75	76	77	77	78	79	80

the age of the subject defines the predicted increment as a function of age. The ability of applicants to meet work output requirements safely is based on the level of heart rate utilization. The cut-off value employed is 75%; applicants falling below 75% are accepted; those between 75% and 80% are classified as marginal, and those above 80% are unacceptable.

Biomechanical function

From the viewpoint of functional capacity evaluation the biomechanical function of significance is musculoskeletal, and the main parameter is that of physical strength.

Physical strength

Muscular strength is the limiting criterion in safe manual materials handling. Strength, however, is a term which has received many definitions and interpretations. Ayoub (1983) points out that it is not a basic quantity such as mass or force, but rather it is an output which depends on the force exerted by a muscle or muscle group in a certain posture, and that in considering strength one must also consider the posture in which that strength was applied. In this connection Kroemer (1975) notes that the relationships of body parts can have substantial effects on the force or torque exerted on an external measuring device. The elbow angle, for instance, substantially determines the amount of force that can be exerted in push–pull with the hand, perpendicular to the axis of the forearm. Textures also greatly influence frictional forces. Surface shapes may greatly affect pressure on the body, causing a pain tolerance limit because of high local pressures.

Strength can be exerted on an object by pushing, pulling, hanging, standing, sitting or leaning. It can be exerted with the limbs or the body, or both. Commonly, however, it is measured by requiring a subject to pull or squeeze on the handle of a dynamometer, which may be an integral electronic device, or may be attached to a chain leading to a floor-mounted load cell or force transducer.

The strength measured can take several forms depending on whether the subject is required to hold, to increase to a maximum, or to jerk. The peak so measured increases in each case (Kroemer, 1975).

While strength, or at least lifting capacity, can be measured by the lifting of simple weights or boxes containing sand or lead shot, various test devices have also been developed.

The devices are basically mechanical or electrical, although pneumatic and hydraulic devices have been used. The mechanical dynamometer, sometimes in the form of a metal frame with two handles which can be squeezed together, is a floor or wall-mounted device connected to a handle which can be grasped and pulled by one or both hands, and is basically a spring the deformation of which is reflected on a calibrated dial and is proportional to the force applied.

Electrical devices commonly adopt the principle of the strain gauge which uses Wheatstone bridge circuitry to measure the deformation in a structure to which force is applied. Like mechanical devices they can be found in many different applications. Their sensitivity is greater than that of the mechancial dynamometers, allowing greater accuracy in measurment.

Interpretation of the results of strength testing requires knowledge and skill. Kroemer (1975) has outlined some of the many variables that can influence the findings, including the nature of the measuring device, the direction of application of force and the location of the point of application relative to the subject, the sex, anthropomorphic form, training and health of the subject, the subject's posture, the method by which the force is exerted (e.g. peak, hold and so on), the motivation of the subject, the environment in which the testing is conducted, as well as the clothing and protective equipment worn. Thus, the simpler the test, and the more standardized the conditions, the more reliable are the results.

As previously noted, the strength that can be exerted is a function of the posture and position of the subject exerting it. In this connection, Chaffin, whose work in biomechanical function is world renowned, assesses the physical stress associated with the exertion of strength as a function of the weight handled and the corresponding strength that can be maintained by an employee or group of employees (Chaffin *et al.*, 1977b). He defines the lift strength ratio (LSR) as:

$$\text{LSR} = \text{weight/position strength}$$

The weight represents the job demand (object handled), and position strength refers to the maximum voluntary static strength that can be exerted at a point in the workplace envelope at which the weight is maintained. The weight to be lifted (or the strength required) should be measured with the position of the hands and body identical with those on the job. Chaffin *et al.* (1977a) outline three methods for the prediction of position strength as follows:

(a) Use values generated from many previous studies, as indicated in Figure C.6.
(b) If skills and resources are available use computer modelling procedures to develop a human model in which torques at articulation joints are within acceptable limits, and angles of body segments are feasible while maintaining balance. The principles of such modelling are outlined in the *Work Practices Guide for Manual Lifting* (AIHA, 1983).
(c) Use an empirically developed predictive formula such as the following:

$$\begin{aligned}\text{Position strength} = 44.177 &+ 0.102 \text{ (arm strength)}\\ &+ 0.0023 \text{ (back strength} \times \text{arm strength)}\\ &+ 2.2245 \text{ (worker height)}\\ &+ 0.6533 \text{ (worker age)}\end{aligned}$$

Chaffin and his colleagues (1977a) showed that the lower the LSR the less the job hazard, and specifically that a value of 0·5 is good for many lifting jobs, or in other words a worker should not be expected to handle more than 50% of his/her static strength, or perhaps 60 lbs (27 kg), a figure which is low by many other standards such as that of the International Labour Office (ILO, 1962, see Table 4.11).

For rating jobs, Chaffin and his colleagues (1977b) propose the following ratios:

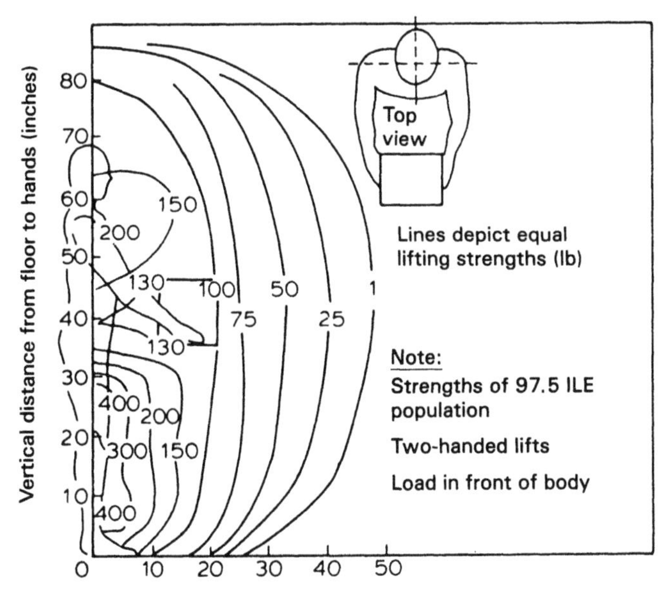

Figure C.6. Predicted strength profile for large strong male (Chaffin et al., 1977)

Job strength rating (JSR)

$$\text{JSR} = \frac{\text{object weight on job}}{\text{average strength of employees on job}}$$

Predicted mean job strength rating (Pred. MJSR)

$$\text{Pred. MJSR} = \frac{\text{object weight on job}}{\text{average predicted strength of employees on job}}$$

For employee matching they defined the following:

Employee strength rating (ESR)

$$\text{ESR} = \frac{\text{object weight on job}}{\text{employee strength demonstrated in job position.}}$$

Psychophysiological method

Rather than attempting the measurement of physical strength, other researchers have examined the human capacity to lift under controlled circumstances. Much of this work has been conducted by Snook and his colleagues (Snook and Ciriello, 1974;

Ciriello and Snook, 1983; Snook, 1978; and summarized by Snook, 1985a,b), as well as by Ayoub and his colleagues, summarized in Ayoub *et al.* (1978).

In Snook's series of lifting tests the subjects lifted boxes with handles (tote boxes) of varying size and weight, at variable frequencies, heights and levels. Subjects also conducted controlled pushing and pulling tasks. The subjects were instructed to adjust the amount of weight or force until it represented the maximum amount that they could perform without becoming tired, weakened, overheated or out of breath. The timing of the test sessions and the environment were controlled. Heart rates and oxygen consumption were also measured.

The results showed that size (specifically width), distance, height and frequency are significant variables to consider when establishing guidelines for maximum acceptable lift. The results also indicate that the maximum acceptable weights and forces for female workers are significantly, but proportionately, similar to the maximum acceptable weights and forces for male workers. The responses to cardiovascular strain are also similar for the two sexes. Indeed, in a review of nine studies of strength capabilities among various populations, Laubach (1976) found that women demonstrated only about 64% of the strength demonstrated by men.

Appendix D
Physical demands and work capacity

The following forms constitute records of a modified version of the Hanman method of evaluation of physical demands and work capacity. They are derived from earlier models developed by Luchterhand and Sydiaha (1966) and Schouppe and Couch (1988). The Physical Demands/Physical Capacities Report Form is used by the job analyst to define the demands of the job, and the same form is used by the medical evaluator to define the capacities of the worker. The Working Conditions Report is completed by the job analyst to define the conditions in which the work is conducted.

When the Physical Demands/Physical Capacities Form is used as a report on physical demands the term Physical Capacities is deleted. Items 1–4 are then completed by inserting in the relevant box the number of hours, or fractions of an hour (minimum $\frac{1}{4}$), that the activity is required during a normal working day (up to 8 h). Items 5–8 are completed by inserting a check mark in any box where there is a demand for normal capacity. Where a task can be completed with limited capacity in some particular area of demand this is indicated by the letter L, while no requirement is indicated by the number 0. The nature of any limited capacity, or other comments, is amplified in the boxes labelled 'Comments'.

When the form is used as a report on Physical Capacities the term Physical Demands is deleted. For items 1–4 a check mark in any box indicates that the worker has normal capacity for any work requiring that activity. Any limitation is expressed by use of a number representing the hours or fractions of an hour that an activity may be pursued in the course of a normal working day. The number 0 is used to indicate that the activity under consideration is not permitted. For items 4–8, the presence, limitation or absence of the capacity is indicated by a check mark, a letter L, or the number 0, respectively. Limitations are amplified in the space provided.

Physical Demands/Physical Capacities Report

Symbols:

Check mark : normal demand/normal capacity
Letter L : limited demand/limited capacity
Number 0 : no demand/no capacity
Numbers : number of hours of demand or capacity

Name ______
Date of birth: ______ Height ______ Weight ______

PHYSICAL CAPACITIES

COMMENTS

Standing ______
Walking ______
Running ______
Sitting

Lifting (kg), includes pushing and pulling effort while stationary

1–2 ______
3–4 ______
5–10 ______
11–20 ______
21–50 ______
50+

Carrying (kg), includes pushing and pulling effort while moving

1–2 ______
3–4 ______
5–10 ______
11–20 ______
21–50 ______
50+

Climbing: legs only ______
legs and arms ______
Balancing ______
Jumping ______
Stooping ______
Kneeling ______
Crawling ______
Crouching ______
Reclining ______
Twisting ______

Feeling R ______
L ______
Fingering R ______
L ______
Handling R ______
L ______
Reaching: Below shoulders R ______
L ______
Above shoulders R ______
L ______

Communication: Speaking ______
Writing ______
Signing ______

Vision

Near ____________________

Far ____________________

Depth ____________________

Colour ____________________

Hearing intelligible sound:

1 m ____________________

3 m ____________________

5 m ____________________

Other sensing:

Taste ____________________

Smell ____________________

Other ____________________

Working Conditions Report

Conditions	Comments
Inside work	
Outside work	
Thermal conditions	
Hot	
Cold	
Wet	
Dry	
Sudden changes	
Atmospheric conditions	
Poor ventilation	
Fumes	
Odours	
Dusts	
Mists	
Gases	
Particulates	
Barometric pressure	
High	
Low	
Hazards	
Noise	
Vibration	

Mechanical
Electrical
Burns
Physical
Chemical
Ionizing energy
Non-ionizing energy
Sharp objects
Falls
Congested worksite
Slippery surfaces
High places
Interpersonal conditions
Working with others
Working around others
Working alone
Working independently but in a group
Interacting with public
Other stresses
Pace of work
Shift work
Overtime
Deadlines
Monotony
Travel

Appendix E
GULHEMP Scale

The following is a summary of the recommendations for different levels of the GULHEMP scale. The full recommendations, with clarification and amplification where required, are presented in the manual developed by Koyl (1974) on behalf of the National Council on the Aging, Inc. It should be emphasized that assignment of a GULHEMP rating, even if carefully determined, is merely a convenient method of recording information. The rating alone does not constitute an assessment of the capacity of a worker, and particularly a handicapped worker, to perform a task. That assessment requires a careful evaluation of the individual in the light of the demands of the task. The GULHEMP scale assists in making the assessment more systematic.

Although a summary is given below, the scales should be used in conjunction with the Manual developed by Koyl (1974) for the National Council on the Aging, Inc.

General physique

G1 *Fit for heavy manual work including digging, lifting, climbing, regularly as main occupation*

Can: Lift maximum — own weight usual — 150 lbs @ 1 per 2 hour — 50 lbs at 10 per hour

: Stand or walk more than 8 hours

: Carry maximum loads short distances (ie. 100 yds) 50 lbs for 50 minutes per hour

: Run, jump, or climb — run or trot 10–20 minutes per hour. Climb 50 minutes per hour

: Climb and carry — could climb 1 flight (18 feet) with maximum load 50 lbs up to 50 minutes per hour

: Forward flexion — 90°, i.e unrestricted

: Environmental restrictions — nil

G2 *Fit for manual work including incidental or occasional heavy work as in G1 — Can change work shifts*

Can: Lift maximum — own weight usual — 150 lbs @ 2 per 8 hours — 25–50 lbs @ 5–6 per hour

: Stand or walk 8 hours

: Carry maximum loads short distances (i.e. 100 yds) 50 lbs for 30–50 minutes per hour

: Run, jump or climb — run or trot 5–10 minutes per hour. Climb 20 flights of 18 feet vertical distance per shift or equivalent on ladders

: Climb and carry usual loads up to 30 minutes out of the hour
: Forward flexion — 90°, i.e. unrestricted
: Environmental restrictions — nil

G3 *Fit for all employment except heavy labour, liable to deteriorate if meals are irregular, or if rest inadequate as with frequent shift changes*
Can: Lift maximum — 100 lbs usual — 10–25 lbs @ 5–10 per hour
: Stand or walk 7 hours — (i.e. most of the shift)
: Carry 25 lbs for 30–50 minutes per hour
: Climb 2 flights per 8 hours
: Climb and carry usual loads 1 flight (18 feet) twice in 8 hours
: Forward flexion — 90° @ 1 per hour. Bends less than 50° — unlimited frequency
: Environmental restrictions — extreme cold as in outside work in cold winter

G4 *Fit for sedentary employment with regular hours and meals*
Can: Lift maximum — 25 lbs @ 1 per hour usual — 5 lbs @ 5–10 per hour
: Stand or walk 2 hours — i.e. sits most of shift
: Run, jump, climb — nil
: Forward flexion — nil
: Environmental restrictions — extreme heat or cold, industrial respiratory allergens

G5 *Fit for restricted or part-time employment*
Can: Lift maximum — 5 lbs usual — 5 lbs
: Stand or walk 1 hour of 2–6 hour day
: Run, jump, climb — nil
: Forward flexion — nil
: Environmental restrictions — cannot work a normal day in any environment

G6 *Fit for self-care only*
Can: Lift 1 lb @ 1 per hour
: 15 minutes of 2–6 hour day
: Environmental restrictions — confined to bed or wheelchair

G7 *Bedfast — unable to care for self*

Upper Extremeties

U1 *Fit to lift strongly to and above shoulder level, to dig, push or drag strongly as a main occupation, e.g. can drive heavy vehicle such as CAT*
Can: Lift (one arm) maximum — 100 lbs @ 1–2 per hour usual — 75 lbs @ 1–2 per hour — 25 lbs @ 10 per hour
: Carry (one arm) 100 lbs for 100 yds @ 1–2 per hour — 25 lbs 50 minutes per hour

: Mobility — 18° shoulders, 170° elbows, 90° forearm, 90° wrist, fingers and thumb straight to press firmly against palm
: Amputations — nil

U2 *Fit to lift strongly to and above shoulder level, to dig, push or drag strongly: fit for manual work; fit for heavy work as U1 incidentally or occasionally*
Can: Lift (one arm) maximum — 85 lbs @ 4 per 8 hours usual — 75 lbs @ 2 per 8 hours — 25 lbs @ 10 per hour
: Carry (one arm) maximum — 75 lbs @ 2 per 8 hour up to 100 yds — 25 lbs, 30–50 minutes per hour
: Bend — 180° shoulders, 170° elbows, 90° forearm, 90° wrist, fingers and thumb straight to press firmly against palm
: Amputations — nil

U3 *Fit for moderately heavy lifting and loading, e.g. can drive light trucks and automobiles*
Can: Lift (one arm) maximum — 75 lbs @ 1–2 per hour usual — 10–15 lbs @ 10 per hour
: Carry (one arm) — 10–15 lbs 30–50 minutes per hour
: Mobility — unilateral restrictions only, e.g. poor arm can bend shoulder 90°, elbow 155°, forearm 135°, wrist (normal) 90°, fingers (normal) straight to press firmly against palm
: Amputations — 4th and 5th fingers off non-dominant hand

U4 *Unilateral disability allowing efficient sedentary or clerical work or light labour*
Better arm — same as U3
Poor arm — can hold or carry light objects such as files, small parts
Can fix work in position to be worked on by good arm
Amputations — unilateral below shoulder

U5 *Bilateral disability or a complete unilateral disability allowing only a few gross or relatively ineffective movements and permitting restricted or part-time employment (the handicapped worker)*
Can: Pick up small parts and files slowly with better arm and work a 4–6 hour day at reduced productivity
: Amputations: bilateral above elbow

U6 *Can give partial self-care* (i.e. may be able to feed self)

U7 *Unable to help self*

Lower extremities

L1 *Able to run, climb, jump, dig and push with sustained effort as a main occupation (e.g. can drive heavy tractors)*
Can: Lift maximum — own weight from squat on one leg plus own weight from squat with two legs usual — 150 lbs @ 1 per 2 hours — 50 lbs at @ 10 per hour

: Carry — same as usual lifts — 150 lbs short distances (i.e. 100 yds) — 50 lbs for 50 minutes per hour
: Climb and carry — 100 lbs 30 minutes per hour — 50 lbs 50 minutes per hour
: Bend — hip 170°, knee 170°, ankle 45°, toes 90°
: Amputations — nil

L2 *Fit for heavy labour. Able to stand, run, climb, jump, push as in L1 incidentally or occasionally*
Can: Lift maximum — same as in L1 usual — 150 lbs @ 2 per 8 hours — 50 lbs @ 5–6 per hour
: Carry — same as usual lifts — 150 lbs short distances (i.e. 100 yds) — 50 lbs for 30–50 minutes per hour
: Bend — same as in L1
: Amputations — all toes in one foot if hard shoes can be worn in trade concerned

L3 *Fit for moderately heavy labour including pushing and digging. Liable to tire if too long on feet (e.g. can drive light vehicles)*
Can: Lift maximum — can come up from squat using best leg only, all available means — 100 lbs @ 1–2 per hour usual — 25 lbs @ 5–6 per hour
: Carry maximum — 100 lbs short distances only usual — 25 lbs @ 30–50 minutes per hour
: Stand — 7–8 hours
: Bend — unilateral restrictions only, e.g. poor leg can bend hip 0°, or ankle 0°, or toes 0°
: Amputations — unilateral below knee with good prosthesis

L4 *Severe unilateral disability or lesser bilateral disability allowing efficient sedentary work or light labour*
Can: Lift maximum — 25 lbs @ 1–2 per hour usual — 5 lbs @ 5–10 per hour
: Carry maximum — 25 lbs at 2 per shift for short distances only usual — 5 lbs @ 5–10 minutes per hour
: Stand — 2 hours
: Bend — (right hip, knee, or ankle) unilateral nil
: Amputations — unilateral, above knee bilateral, below knee

L5 *Bilateral or severe unilateral disability allowing a few relatively ineffective movements and permitting restricted employment. Fit for sedentary work only*
Can: Walk with use of crutches or canes
: Amputations — bilateral above knee

L6 *Unable to accept employment because of severity of disability*

L7 *Bedfast*

Hearing

H1 Able to hear well enough for any employment
H2 Able to hear well enough to be employed anywhere that acuity of hearing is not the reason for employment
H3 Able to hear well enough to be employed where slight degrees of hearing loss are unimportant.
H4 Able to hear well enough to work where severe impairment of hearing is not a handicap.
H5 Functionally completely deaf, but with no additional symptoms and can read lips
H6 Functionally deaf with progressive disease, tinnitus and imperfect lip reading
H7 Completely deaf with active disease and unable to read lips

Eyesight

E1 Able to see sufficiently well with or without glasses for any employment including work where good vision is the reason for employment. Night vision and colour vision normal
E2 Able to see sufficiently well with corrective glasses for any employment where good vision is the reason for employment
E3 Able to see well enough with one eye to do work not requiring binocular vision
E4 Able to see well enough with one eye with glasses to do work, except close work. No rapidly progressing disease
E5 Able to see well enough to get around with glasses and do ordinary labour
E6 Able to see shapes dimly, or blind and trained
E7 Severe progressive disease, or blind and untrained

Mental

M1 Excellent language skills, oral and written; flexible, ingenious problem-solving; advanced (and appropriate) education; leadership skill and experience. IQ 130 and above
M2 Very good language skills, oral and written; flexible, ingenious problem solving; more than average education and capable of acquiring high-level training appropriate to the job. IQ 110–129
M3 Average language skills; average education; capable of learning average job quickly. IQ 90–109
M4 Able to read and write routine material; able to learn simple routine jobs; may appear to be deteriorating intellectually; somewhat difficult to retrain or reallocate. IQ 80–98

M5 Oral and/or written language handicap; very limited literacy; definitely deteriorating intellectually (e.g. very poor memory); very limited capacity for learning and retraining. IQ 70–79

M6 Severe communication defect, such as severe speech or language handicap; severe specific learning disability; almost total illiteracy. Employable only under unusual circumstances because of handicaps in communication or learning capacity. IQ 60–69

M7 Totally incapacitating mental handicaps or communication handicaps; unemployable in any capacity unless handicaps are remedial. IQ 59 or below

Personality

P1 Stable predictable behaviour. Able to use intellectual gifts to make decisions quickly and rationally. Realistic self-esteem. Good judgment in making logical decisions and in dealing with other people. Vigorous drive for achievement. Able to stimulate employees to do their best

P2 As in P1, above, but may have some minor problems in productivity and/or interpersonal relationships which limit possibilities of advancement to some degree. Can take direction equably when it is appropriate

P3 In general, reliable and consistent. Takes responsibility well, but mainly for own work, rather than in a supervisory capacity. Has qualities of personality and character which limit prospects of promotion to a considerable degree. (This is the category of the 'average' employee)

P4 Needs encouragement and/or guidance to function effectively. Does not take responsibility well, overreacts to stress, causes trouble among fellow employees at times

P5 Needs much encouragement, guidance and supervision. Overwhelmed by unusual stresses, does not adapt well to change, works productively only in familiar surroundings with protective supervision

P6 Frequent psychosomatic and/or emotional breakdowns. Frequent severe conflicts with other employees. Accomplishes little work. Uses up most of his/her energy in self-defeat or troublemaking. Severe character defects

P7 Unemployable in any capabity due to overt mental illness

Record of rating

The GULHEMP rating represents the state of fitness of the person at the time of the examination. It may be presented for record purposes as shown in Table E.1.

Table E.1. Record of GULHEMP rating (after Koyl, 1974).

Name: JONES, John I.D. NUMBER: 2360

Date of examination: May 18 1962

Present status	Y.O.B.	G U L H E M P
	62	2 1 1 2 2 3 3
Expected change in	Years	
	5	nil

Remarks: Mild allergy to acrylic resins with prolonged contact. Should not be transferred to plastics shop without reference to Medical Department

Signed:____________________ M.D.

Suffixes

Suffixes may be added to the GULHEMP profile to reduce the necessity for cumbersome and repetitious statements in a 'Remarks' block. For example, the preceding problem with allergy to acrylics might be sufficiently common as to warrant a 'P' suffix thus:

Y.O.B.	G U L H E M P
62	2 1 1 2 2 3 3 P
Remarks:	nil

Should a condition be deemed to be remedial an 'R' suffix might be added to the numerical rating in the appropriate position. Thus an otherwise fit employee with a broken arm rendering him temporarily unfit for full activity might be awarded a temporary U4 rating, thus:

G	U	L	H	E	M	P
2	4R	1	1	2	3	3

Other suffixes can be developed as the need arises, but it is emphasized that suffixes and similar modifications should be kept to a minimum to preserve the simplicity.

Job matching (GULHEMP)

For job matching purposess it is necessary to set up GULHEMP profiles for the job or tasks under consideration. It must be recognized that such profiles are the minimum

Table E.2. GULHEMP profiles.

Job	GULHEMP profile						
Electronic technician	2	3	2	2	2	3	3
Electronic mechanic	2	3	2	2	2	3	3
Electrical mechanic (shop)	2	3	2	3	2	3	3
Electrical mechanic (line)	2	3	1	3	2	3	3
Radar installer	2	2	2	3	2	3	3
Electrical installer (shop)	2	3	3	3	2	3	3
Electrical installer (line)	2	2	2	3	2	3	3
Photographic mechanic	4	3	4	4	2	3	3
Instrument technician	2	2	2	2	2	3	3
Instrument mechanic	2	3	3	3	2	3	3
Instrument assembler	2	3	3	4	2	3	3
Clerk — general duty	3	4	4	4	3	4	4
Clerk — library engineering	4	4	4	4	4	4	4
Clerk — information systems	3	3	2	4	4	4	4
Clerk — data processing	4	4	4	4	3	3	3

acceptable profiles, established by determining the minimal level of fitness required of a worker at that job in each of the rating categories.

To establish a match none of the numbers in the worker's profile may be below those of the job profile in the same category. Under most circumstances this may be all that is required. In the case of the handicapped worker, however, it would normally be desirable to use the apparent match or lack of match as a guideline indicating the need for closer and individual examination in the light of a clear understanding of the worker's capacities and the demands of the job.

Table E.2 gives, as an illustration, some GULHEMP profiles for selected positions in an aircraft factory (Koyl, 1974).

References and bibliography

Abt Associates, 1984, *Job Matching: The Assessment of Individuals for Work*, Handicapped Employment Program, Ontario Ministry of Labour, Canada.

AIHA, 1983, *Work Practices Guide for Manual Lifting*, NIOSH Publication 81–122, Cincinnati, OH: National Institute for Occupation Safety and Health, United States Public Health Service.

Alexander, R. W., Maida, A. S. and Walker, R. J., 1975, The validity of pre-employment examinations, *Journal of Occupational Medicine*, **17**, 687–92.

AMA, 1964, Employment of the handicapped. Council on Occupational Health, *American Medical Association Journal*, **187**, 235.

AMA, 1968, Employability of workers handicapped by certain diseases, *Archives of Environmental Health*, **17**, 1–9.

AMA, 1969, *Medical Aspects of Automotive Safety*, Chicago, IL: American Medical Association (reviewed 1972).

AMA, 1973, *Grinding principles of medical examinations in industry*, Chicago, IL: Department of Environmental, Public, and Occupational Health, Division of Scientific Activities.

Anderson, C. K., 1987, Pre-placement screening: survival of the fittest, *Risk Management*, **34**, 44–6.

Anon., 1976, *Basic Principles of Vocational Rehabilitation of the Disabled*, Geneva: International Labour Office.

Anon., 1978, *Adaptation of Jobs for the Disabled*, Geneva: International Labour Office.

ANSI, 1960, *Document S3.1-1960*, New York: American National Standards Institute.

Armstrong, T. J., Foulke, J. A., Joseph, B. S. and Goldstein, S. A., 1982, Investigation of cumulative trauma disorders in a poultry processing plant, *American Industrial Hygiene Association Journal*, **43**, 103–15.

Astrand, I., 1960, Aerobic work capacity in men and women with special reference to age, *Acta Physiologica Scandinavica*, **49** (Suppl. 169).

Astrand, P.-O. and Rhyming, I., 1954, A nomogram for calculation of aerobic capacity (physical fitness) from pulse rate during submaximal work, *Journal of Applied Physiology*, **7**, 218.

Astrand, P.-O. and Rodahl, K., 1970, *Textbook of Work Physiology*, New York: McGraw-Hill.

Ayoub, M. A., 1971, 'A biomechanical model for the upper extremity using optimization techniques', unpublished PhD thesis, Texas Technical University.

Ayoub, M. A., 1982, Pre-employment screening programs that match job demands with worker abilities — Part 2, *Industrial Engineering*, **14**, 41–6.

Ayoub, M. A., 1983, Design of a pre-employment screening program, in *Ergonomics of Work Station Design*, edited by T. O Kvalseth, London: Butterworths, pp. 152–85.

Ayoub, M. M., 1972, Human movement recording for biomechanical analysis, *International Journal of Production Research*, **10**, 35–51.

Ayoub, M. M., Dryden, R. D., McDaniel, J. W., Knipfer, R. E. and Aghazadeh, F., 1978, Modeling of lifting capacity as a function of operator and task variables, in *Safety in Manual Materials Handling*, edited by C. G. Drury, NIOSH Publication 78–185, Cincinnati, OH: National Institute for Occupational Safety and Health, United States Public Health Service, pp. 120–30.

Ayoub, M. M., Dryden, R., McDaniel, J., Knipfer, R. and Dixon, D., 1979, Predicting lifting capacity, *American Industrial Hygiene Association Journal*, **40**, 1075–84.

Baleshta, M. M., 1986, 'A movement notation system for ergonomic analysis', unpublished MSc thesis, University of Waterloo, Canada.

Baleshta, M. M. and Fraser, T. M., 1986, An arm movement notation system, in *Trends in Ergonomics/Human Factors III, Part B*, edited by W. Karwowski, Amsterdam: North Holland, pp. 613–21.

Barrett, G. V., Dambrott, F. H. and Smith, G. R., 1975, Relationships between individual attributes and job design, *Document No 0014-75-0202-001, NR 1S1-351*, Washington, DC: United States Office of Naval Research.

Bartley, D. L., 1981, *Job Evaluation: Wage and Salary Administration*, Don Mills, Ontario: Addison-Wesley.

Bauman, M. K. and Hayes, S. P., 1951, *A Manual for the Psychological Examination of the Adult Blind*, New York: Psychological Corporation.

Becker, W. F., 1955, Prevention of back injuries through pre-employment examination, *Industrial Medicine and Surgery*, **24**, 486.

Bernauer, E. M. and Bonanno, J., 1975, Development of physical profiles for specific jobs, *Journal of Occupational Medicine*, **17**, 27–33.

Berwitz, C., 1975, *The Job Analysis Approach to Affirmative Action*, New York: John Wiley.

Bigos, S. J. and Battie, M. C., 1987, Pre-placement worker testing and selection considerations, *Ergonomics*, **30**, 249–51.

Bond, M. B., 1975, Health services for small businesses, in *Occupational Medicine*, edited by Carl Zenz, Chicago, London: Year Book Medical Publishing, pp. 3–19.

Borg, G., Dahlstrom, H., 1960, *The Perception of Muscle Work*, Sweden: Umea Research Library.

Borg, G. A. V., 1962, *Physical Performance and Perceived Exertion*, Copenhagen: Ejnar Munskgaard.

Bosco, M. L., 1985, Job analysis, job evaluation, and job classification, *Personnel*, **62**, 70–4.

Brooklyn, P. L., Preemployment testing, *Personnel*, **65**, 66–8.

Brouha, L., 1960, *Physiology in Industry. Evaluation of Industrial Work Stress by the Physiological Reactions of the Worker*, Oxford: Pergamon Press.

Brubaker, W. W., 1972, Pre-employment physical examinations, *Pennsylvania Medicine*, **75**, 53–6.

Bullis, M. and Foss, G., 1986, Assessing the employment-related interpersonal competence of mildly mentally retarded workers, *American Journal of Mental Deficiency*, **91**, 43–50.

Campion, M. A., 1983, Personnel selection for physically demanding jobs: review and recommendations, *Personnel Psychology*, **36**, 527–50.

Campione, K. M., 1972, The pre-employment examination: an evaluation, *Industrial Medicine and Surgery*, **41**, 27–30.

Canadian Medical Association, 1988, *Provision of Occupational Health Services: A Guide for Physicians*, Ottawa, Canada: Department of Communications and Government Relations, Canadian Medical Association.

Caplan, R. D., Cobb, S., French, J. R. P., Van Harrison, R. and Pinneau, S. W.Jr, 1975, *Job Demands and Workers' Health*, Cincinnati, OH: National Institute for Occupational Health and Safety, United States Public Health Service.

Catalina, P., 1983, Medical examination of workers, in *Encyclopedia of Occupational Health and Safety*, edited by L. Parmeggiani, Geneva: International Labour Office, pp. 1308–10.

Chaffin, D. B., 1975, Biomechanics of manual materials handling and low-back pains in

Occupational Medicine, edited by Carl Zenz, Chicago: Year Book Publishers, pp. 443–67.
Chaffin, D. B., Herrin, G. D., Keyserling, W. M. and Foulke, J. A., 1977a, *Pre-employment Strength Testing*, NIOSH Publication No. 77-163, Cincinnati, OH: National Institute for Occupational Safety and Health, United States Public Health Service.
Chaffin, D. B., Herrin, G. D., Keyserling, W. M. and Foulke, J. A., 1978, Pre-employment strength testing — an updated position, *Journal of Occupational Medicine*, **20**, 403–8.
Chaffin, D. B., Herrin, G. D., Keyserling, W. M. and Garg, A., 1977b, A method for evaluating the biomechanical stresses resulting from manual materials handling jobs, *American Industrial Hygiene Association Journal*, **38**, 662–75.
Ciriello, V. M. and Snook, S. H., 1983, A study of size, distance, height and frequency effects on manual handling jobs, *Human Factors*, **35**, 473–83.
Cohen, V. and Pfeffer, J., 1986, Organizational hiring standards, *Administrative Science Quarterly*, **31**, 1–24.
Colombini, D., Occhipinti, E., Moetine, G., Grieco, A., Pedotti, A., Boccardi, S., Frigo, C and Menoni, O., 1985, Posture analysis, *Ergonomics*, **28**, 275–84.
Cooper, W. F., 1980, The techniques of selective placement of factory employees with major physical disabilities, *Australian Family Physician*, **9**, 109–14.
Corlett, E. N., Madeley, S. J. and Manenica, I., 1979, Posture targetting: a technique for recording working postures, *Ergonomics*, **22**, 357–66.
Cowell, J. W., 1986, Guidelines for fitness-to-work examinations, *Canadian Medical Association Journal*, **135**, 985–8.
Crookshank, J. W. and Warshaw, L. M., 1961, Detecting potential low-back disabilities, *Southern Medical Journal*, **54**, 636.
Daniel, C., 1986, Science, system, or hunch: alternative to improving employee selection, *Public Personnel Management*, **15**, 1–10.
Davey, B. W., 1984, Personnel testing and the search for alternatives, *Public Personnel Management*, **13**, 361–74.
David, G. C., 1985, Intra-abdominal pressure measurements and load capacities for females, *Ergonomics*, **28**, 345–88.
Dulevicz, V., 1984, Uses and abuses of selection tests, *Personnel Management*, **16**, 46–7.
Dunnette, M., 1982, Critical concepts in the assessment of human capabilities, in *Human Performance and Productivity*, Vol. 1, *Human Capability Assessment*, edited by M. Dunnette and E. A. Fleishman, Hillsdale, NJ: Erlbaum, pp. 1–11.
Durnin, J. and Passmore, R., 1967, *Energy, Work and Lesiure*, London: William Heineman.
Eisler, H., 1962, Subjective scale of force for a large muscle group, *Experimental Psychology*, **64**, 253–7.
Eskol, M., and Wachmann, A., 1958, *Movement Notation*, London: Weidenfield and Nicholson.
Farina, A. J. Jr, 1969, Development of a taxonomy of human performance. A review of descriptive schemes for human task behavior, *Report No. 126 1/69-TR2*, Silver Springs, MD: American Institute for Research.
Feild, H. S. and Gatewood, R. D., 1987, Matching talent with the task, *Personnel Administrator*, **32**, 113–4.
Fine, S. A., 1973, *Functional Job Analysis Scales: A Desk Aid*, Methods for Manpower Analysis, Number 7, Kalamazoo, MI: W. E. Upjohn Institute for Employment Research.
Fine, S. A., 1974, *Functional Job Analysis Scales*, Methods for Manpower Analysis, No 7, Kalamazoo, MI: W. E. Upjohn Institute for Employment Research.

Fine, S. A., Functional job analysis: an approach to a technology for manpower planning, *Personnel Journal*, November, 813–8.

Fine, S. A. and Wiley, W. W., 1971, *An Introduction to Functional Job Analysis*, Kalamazoo, MI: W. E. Upjohn Institute for Employment Research.

Fine, S. A., Holt, A. M. and Hutchinson, M. F., 1974, Functional job analysis: how to standardize task statements, *Methods for Manpower Analysis, No. 9*, Kalamazoo, MI: W. E. Upjohn Institute for Employment Research.

Flanagan, J. C., 1954, Critical incident technique, *Psychological Bulletin*, **51**, 327–58.

Fleishman, E. A., 1954, Dimensional analysis of psychomotor abilities, *Journal of Experimental Psychology*, **48**, 437–54.

Fleishman, E. A., 1962, The description and prediction of perceptual motor skill training, in *Training Research and Education*, edited by R. Glaser, Pittsburgh: University of Pittsburgh Press.

Fleishman, E. A., 1964, *The Structure and Measurement of Physical Fitness*, Englewood Cliffs, NJ: Prentice-Hall.

Fleishman, E. A., 1975, Towards a taxonomy of human performance, *American Psychologist*, **30**, 1127–49.

Fleishman, E. A., 1979, Evaluating physical abilities required by jobs, *Personnel Administration*, **24**, 82–90.

Fleishman, E. A., 1988, Some new frontiers in personnel selection research, *Personnel Psychology*, **411**, 679–701.

Fleishman, E. A. and Hogan, J. C., 1984, A taxonomic method for assessing the physical requirements of jobs: the physical abilities approach, in *Medical Standards Project Final Report*, 3rd Ed, edited by S. W. Nylander and G. Carmean, San Bernardino, CA: Office of Personnel Management, pp. 29–49.

Fleishman, E. A., Gebhardt, D. L. and Hogan, J. C., 1984, The measurement of effort, *Ergonomics*, **27**, 947–54.

Fleishman, E. A., Kremer, E. J. and Shoup, G. W., 1961a, The dimensions of physical fitness — a factor analysis of strength tests. *Technical Report No. 2*, Contract Nonr 609 (32), Yale University, NY: Office of Naval Research.

Fleishman, E. A., Thomas, P. and Munroe, P., 1961b, The dimensions of physical fitness — a factor analysis of speed, flexibility, balance and coordination tests. *Technical Report No. 3*, Contract Nonr 609 (32), Yale University, NY: Office of Naval Research.

Frank, M. S., 1982, Position classification: a state-of-the-art review and analysis, *Public Personnel Management*, **11**, 239–47.

Fraser, T. M., 1977, An approach to the quantitative analysis of human reliability. *Proceedings of the 4th Meeting of the Human Factors Working Group of the International Air Transport Association*, Ames, CA: National Aeronautics and Space Agency.

Fraser, T. M., 1989, *The Worker at Work*, London: Taylor & Francis.

Fraser, T. M. and Samuel, B., 1979, A systems ergonomic approach towards matching worker capacity and job demand. *Proceedings of the 7th Congress of the International Ergonomics Association*, Warsaw, Poland.

French, J. W., 1951, The description of aptitude and achievement tests in terms of rotated factors, *Psychometric Monographs*, No. 5.

French, J. W., Ekstrom, R. B. and Price, L. A., 1963, *Manual for Kit of Reference Tests for Cognitive Factors*, Princeton, NJ: Educational Testing Services.

Fried, Y. and Ferris, G. R., 1987, The validity of the job characteristics model: a review and meta-analysis, *Personnel Psychology*, Summary V40, 287–322.

Gael, S., 1983, *Job Analysis. A Guide to Assessing Work Activities*, San Francisco: Josey-Bass.

Gamberale, F., 1985, The perception of exertion, *Ergonomics*, **28**, 299–308.

Gamberale, F., Holmer, I., Kindblom, A-S. and Nordstrom, A., 1978, Magnitude perception of added inspiratory resistance during steady state exercise, *Ergonomics*, **21**, 531–8.

Gebhardt, D. L., Jennings, M. C. and Fleishman, E. A., 1981, Factors affecting the reliability of physical ability and effort ratings of Navy tasks. *Technical Report No. 3034/R81-1*, Washington, DC: Advanced Research Resources Organization.

Gibson, R. L., 1974, Why the preplacement physical isn't out of date, *International Journal of Occupational Health and Safety*, **43**, 31–3.

Goldman, R. H., 1986, General occupational health history and examination, *Journal of Occupational Medicine*, **28**, 967–74.

Gordon, G., 1975, *The Application of GPSS (General Purpose Simulation System) to Discrete System Simulation*, Englewood Cliffs, NJ: Prentice-Hall.

Guildford, J. P., 1967, *The Nature of Human Intelligence*, New York: McGraw-Hill.

Guildford, J. P. and Hoepfner, R., 1971, *The Analysis of Intelligence*, New York: McGraw-Hill.

Halperin, W. E., Ratcliffe, J. M. and Frazier, T. M., 1984, Medical screening in the workplace: proposed principles. *Conference on Medical Screening and Biological Monitoring for the Effects of Exposure in the Workplace.*

Hanks, T. G., 1962, The physical examination in industry: a critique, *Archives of Environmental Health*, **5**, 365–74.

Hanman, B., 1945, Matching the physical characteristics of workers and jobs, *Industrial Medicine*, **14**, 405–30.

Hanman, B., 1946, Placing the handicapped — a positive, individual and specific approach, *Industrial Medicine*, **15**, 597–604.

Hanman, B., 1948, Placement of disabled personnel. *Proceedings of the 9th International Congress on Industrial Medicine*, London.

Hanman, B., 1958, The evaluation of physical ability, *New England Journal of Medicine*, **258**, 986–93.

Harte, J. D., 1974, Is the pre-employment medical examination of value? *Proceedings of the Royal Society of Medicine*, **67**, 177–180.

Hipp, L. L., Jirak, P. D. and Golonka, E. J., 1977, Evaluation of an occupational health examination program, *Journal of Occupational Medicine*, **19**, 205–7.

Hogan, J. C. and Bernacki, E. J., 1981, Developing job-related pre-placement medical examinations, *Journal of Occupational Medicine*, **23**, 469–76.

Hogan, J. C. and Fleishman, E. A., 1979, An index of the physical effort required in task performance, *Journal of Applied Physiology*, **64**, 197–204.

Hogan, J. C. and Quigley, A. M., 1986, Physical standards for employment and the courts, *American Psychologist*, **41**, 1193–217.

Hogan, J. C., Ogden, G. D., Gebhardt, D. L. and Fleishman, E. A., 1980, Reliability and validity of methods for evaluating perceived physical effort, *Journal of Applied Physiology*, **65**, 672–9.

Holzman, P., 1982, ARBAN — a new method for analysis of ergonomic effort, *Applied Ergonomics*, **13**, 82–86.

Human Rights Commission, 1981, *Guidelines for Assessing Accommodation Requirements for Persons with Disabilities*, Toronto, Ontario, Canada: Ministry of Citizenship.

Hutchinson, A., 1977, *Labanotation*, 3rd Edn, New York: Theatre Arts Books.

ILO, 1962, Recommended weight limits for lifting tasks, *CIS Information Sheet, 3*, Geneva: International Labour Office.

Jones, M. and Prien, E. P., 1978, A valid procedure for testing the physical abilities of job applicants, *Personal Administrator*, **23**, 33–8.

Kelly, F. J., 1965, Pre-employment medical examinations including X-rays, *Journal of Occupational Medicine*, **7**, 132–6.

Kinsman, R. A. and Weiser, J., 1976, Subjective symptomatology during work and fatigue, in *Psychological Aspects and Physiological Correlates of Work and Fatigue*, edited by E. Simonson and P. C. Weiser, Springfield, IL: Charles C. Thomas, pp. 336–403.

Koyl, L. F., 1974, *Employing the Older Worker*, New York: National Council on the Aging, Inc.

Kress, A. L., 1939, How to rate jobs, and when, *Factory Management*, **97**, 60–5.

Kroemer, K. H. E., 1970, Human strength: terminology, measurement and interpretation of data, *Human Factors*, **12**, 297–313.

Kroemer, K. H. E., 1975, Measurements of muscular strength capabilities, in *Engineering Anthropometry Methods*, edited by J. A. Roebuck Jr, K. H. E. Kroemer and W. G. Thomson, New York: John Wiley, pp. 108–28.

Kroemer, K. H. E., 1985, Testing individual capability to lift material: repeatability of a dynamic test compared with static testing, *Journal of Safety Research*, **16**, 1–7.

Landau, K., 1978, 'Arbeitswissenschaftliche ehrebungsverfahren zur tatigkeitsanalyse', Dissertationschrift S 17, Technische Hochschule, Darmstadt.

Laubach, L. L., 1976, Comparative muscular strength of men and women: a review of the literature, *Aviation, Space, and Environmental Medicine*, **47**, 534–42.

Liles, D. H., Deivanayagam, S., Ayoub, M. M. and Mahajan, P., 1984, A job severity index for the evaluation and control of lifting injury, *Human Factors*, **26**, 683–93.

Lincoln, T. A., 1968, The role of the occupational physician in the evaluation of applicants for employment, *Southern Medical Journal*, **61**, 359–62.

Livy, B., 1975, *Job Evaluation. A Critical Review*, London: Allen and Unwin.

Lott, M. R., 1926, *Wage Scales and Job Evaluation*, New York: Ronald Press.

Luchterhand, E. and Sydiaha, D., 1966, *Choice in Human Affairs: An Application to Aging — Accident — Illness Problems*, New Haven, CT: College and University Press.

Luongo, E. P., 1962, The pre-placement physical examination in industry: its values, *Archives of Environmental Health*, **5**, 358–64.

Lytel, R. B. and Botterbusch, K. F., 1981, *Physical Demands Job Analysis: A New Approach*, Menomonie, WI: Materials Development Center, Stout Vocational Institute, University of Wisconsin-Stout.

McCormick, E. J., 1979, *Job Analysis: Methods and Applications*, New York: Amacom (A Division of the American Management Association).

McCormick, E. J., Jeanneret, P. R. and Mecham, R. C., 1972, A study of job characteristics and job dimensions as based on the position analysis questionnaire (PAQ), *Journal of Applied Psychology*, **56**, 347–68.

McGill, C. M., 1968, Industrial back problems: a control program, *Journal of Occupational Medicine*, **10**, 174.

McGuinness-Scott, J., 1983, *Movement Study and Benesh Movement Notation*, London: Oxford University Press.

Malzahn, D. E., 1980, An ability evaluation system for persons with physical disabilities. *Proceedings of the 34th Annual Meeting of the Human Factors Society, Santa Monica, CA*.

Markowitz, J., 1987, Managing the job analysis process, *Training and Development Journal*, **41**, 64–6.

Melching, W. H. and Borcher, S. D., 1973, Procedures for constructing and using task inventories, *Center for Vocational and Technical Education, Research and Development Series, No. 91*, Columbus, OH: The Ohio State University.

Melvian, J. G. and Maxwell, R. B., 1974, *Reliability and Maintainability Manual Process System*, Chalk River, Ontario: Atomic Energy of Canada Ltd.

Michaels, I., 1973, A plea for abandonment of the complete history and physical examination, *Canadian Medical Association Journal*, **108**, 299–300.

Ministry of Fitness and Amateur Sport, 1986, *Canadian Standardized Test of Fitness, Operations Manual* 3rd Edn, Document FAS 7378, Government of Canada, Ottawa: Ministry of State for Fitness and Amateur Sport.

Mock, H. E., 1920, *Industrial Medicine and Surgery*, Philadelphia: W. B. Saunders.

Moniz, A. E., 1983, Handicapped and disabled persons, in *Encyclopedia of Occupational Health and Safety*, 3rd Edn, edited by L. Parmeggiani, Geneva: International Labour Office, 994–7.

Morsh, J. E. and Archer, W. B., 1967, *Procedural Guide for Conducting Occupational Surveys in the United States Air Force*, Document No. PRL-TR-67-11, AD-664-36, Lackland AFB, TX: Personnel Research Laboratory, United States Air Force.

Morsh, J. E. and Christabel, R. E., 1966, *Impact of the Computer on Job Analysis in the United States Air Force*, Document No. TR-66-19, Lackland AFB, TX: Personnel Research Laboratory, United States Air Force.

Myklebust, H. R., 1964, *The Psychology of Deafness, Sensory Deprivation, Learning and Adjustment*, 2nd Edn, New York: Grune and Stratton.

Nester, M. A., 1984, Employment testing for handicapped persons, *Public Personnel Management*, **13**, 417–34.

NIOSH, 1981, *Work Practices Guide for Manual Lifting*, Cincinnati, OH: US Department of Health and Human Services, Public Health Service, Centers for Disease Control, National Institute for Occupational Health and Safety, Division of Biomedical and Behavioral Science.

Nordgren, B., 1972, Anthropometric measures and muscle strength in young women, *Scandinavian Journal of Rehabilitation Medicine*, **4**, 165–69.

Nylander, S. W. and Carmean, G., 1984, *Medical Standards Project Final Report*, Vol. 1, 3rd Edn, San Bernardino, CA: Office of Personnel Management, Personnel Division.

Ontario Ministry of Labour, 1981, *Handicapped Employment: Essential Physical Demands Analysis*, Toronto, Ontario, Canada: Handicapped Employment Program.

Pandolf, K. B., Cafarell, E., Noble, B. J. and Metz, K. F., 1972, Perceptual responses during prolonged work, *Perceptual and Motor Skills*, **35**, 975–85.

Patla, A., Eickmeyer, W., Calvert, T. and Rymen, R., 1985, An interactive computer system, *Physiotherapy Canada*, **37**, 230.

Patla, A., Norman, R. W., Clause, S. and Eichmeyer, W., 1984, A tool for description, reading, and storage of ergonomics data, *Proceedings of the 17th Annual Conference of the Human Factors Association of Canada*, Hamilton, Ontario, Canada.

Patrick, J. and Moore A. K., 1985, Development and reliability of a job analysis technique, *Journal of Occupational Psychology*, **58**, 149–58.

Patton, J. A., Littlefield, C. L. and Self, S. A., 1964, *Job Evaluation — Text and Cases*, Homewood, IL: Richard D. Irwin.

Pierce, J. L., McTavish, D. G. and Knudsen, K. R., 1986, The measurement of job characteristics: a content and contextual analytic look at scale viability, *Journal of Occupational Behaviour*, **7**, 299–313.

Present, A. J., 1974, Radiography of the lower back in pre-employment examinations, *Radiology*, **112**, 229–30.

Priel, V., 1974, A numerical definition of posture, *Human Factors*, **16**, 576.

Prien, E. P., Goldstein, I. L. and Macey, W. H., 1987, Multidomain job analysis: procedures and applications, *Training and Development Journal*, **41**, 68–72.

Priest, J. W. and Roessler, R. T., 1983, Job analysis and workplace design resources for rehabilitation, *Rehabilitation Literature*, **44**, 201–5, 216.

Primoff, E. S., 1973, *How to Prepare and Conduct Job-element Examinations*, Technical Study, 75–1, Washington, DC: US Civil Service Commission.

Rahimi, M. and Malzahn, D. E., 1984, Task design and modification based on physical ability measurement, *Human Factors*, **26**, 715–26.

Ramanujam, S., 1988, 'Job matching: physical demands of the job versus functional capacities of the employee, Ontario Human Rights Commission', presentation at The Seminar on Job Matching, Centre for Occupational Health and Safety, University of Waterloo, Ontario.

Rohmert, W., 1985, AET — a new job analysis method, *Ergonomics*, **28**, 245–54.

Rohmert, W. and Landau, K., 1983, *A New Technique for Job Analysis*, London and New York: Taylor & Francis.

Salvendy, G. and Seymour, W., 1973, *Prediction and Development of Industrial Work Performance*, New York: Wiley Interscience.

Samuel, B. and Fraser, T. M., 1979, An ergonomic approach towards matching worker capacity with job demand, *VIIIth Congress of the International Ergonomics Association*, Warsaw, Poland.

Sawyer, W. A., 1942, Prehabilitation and rehabilitation in industry, *Journal of the American Medical Association*, **119**, 419.

Schender, R. F. and McDonough, T. J., 1971, Experience data on university students hired without pre-employment examination, *Industrial Medicine*, **13**, 363–70.

Schilling, R. S. F., 1984, 'Role of medical examination in protecting worker health', presentation at The Conference on Medical Screening and Biological Monitoring in the Workplace, Cincinnati, OH.

Schlei, B. L. and Grossman, P., 1976, *Employment Discrimination Law*, Washington, DC: Bureau of National Affairs.

Schober, H. A. W., 1983, Visual acuity, in *Encyclopedia of Occupational Health and Safety*, 3rd Edn, edited by L. Parmeggiani, Geneva: International Labour Office, pp. 2267–70.

Scholl, C. and Schnur, R., 1976, *Measures of Psychological, Vocational and Educational Functioning in the Blind and Visually Handicapped*, New York: American Foundation for the Blind.

Schouppe, K. and Couch, D. B., 1988, 'Application of job matching, physical requirements of select occupations at Freeport Hospital', presentation at The Seminar on Job Matching, Centre for Occupational Health and Safety, University of Waterloo, Ontario.

Schussler, T., Kaminer, A. J., Power, V. L. and Pomper, I. H., 1975, The pre-placement examination: an analysis, *Journal of Occupational Medicine*, **7**, 254–7.

Shevick, B. H., 1961, The pre-employment examination and the low back syndrome, *Iowa Medical Society*, **51**, 651–4.

Slavenski, L., 1986, Matching people to the job, *Training and Development Journal*, **40**, 54–7.

Smith, J., Smith, L. and McLaughlin, T., 1982, A biomechanical analysis of industrial manual materials handlers, *Ergonomics*, **25**, 299–308.

Snook, S. N., 1978, The design of manual handling tasks, *Ergonomics*, **21**, 963–85.

Snook, S. N., 1985a, Psychophysical considerations in permissible loads, *Ergonomics*, **28**, 329–30.

Snook, S. N., 1985b, Psychophysical acceptability as a constraint in manual working capacity, *Ergonomics*, **28**, 331–5.

Snook, S. H. and Ciriello, V. M., 1974, Maximum weights and workloads acceptable to female workers, *Journal of Occupational Medicine*, **16**, 527–34.

Stephens, S. S., 1957, On the psychophysical law, *Psychological Review*, **64**, 153–81.
Stevens, J. C. and Mack, J. D., 1959, Scales of apparent force, *Journal of Experimental Psychology*, **58**, 405–13.
Stillman, N. G., 1979, The law in conflict: accommodating equal employment and occupational health obligations, *Journal of Occupational Medicine*, **21**, 599–606.
Stone, C. H. and Yoder, D., 1970, *Job Analysis*, Long Beach, CA: California State College.
Stunkel, E. R., 1957, The performance of deaf and hearing college students on verbal and non-verbal intelligence tests, *American Annals of the Deaf*, 342–55.
Swain, A. D., 1973, Design of industrial jobs a worker can and will do, *Human Factors*, **15**, 129.
Terry, D. R. and Evans, R. N., 1973, *Methodological Study for Determining the Task Content of Dental Auxiliary Education Programs*, Bureau of Health Management Education, Document HRP 000-4628, Bethesda, MD: National Institutes of Health.
Theologus, G. C., Romashko, T. and Fleishman, E. A., 1973, Development of a taxonomy of human performance. Validation study of ability scales for classifying human tasks, *Catalog of Selected Documents in Psychology*, **3**, 29 (MS no. 326).
Thompson, D. E. and Thompson, T. A., 1982, Court standards for job analysis in test validation, *Personnel Psychology*, **35**, 865–74.
Todd, J. W., 1965, Pre-employment medical examination, *Lancet*, **i**, 797–9.
Ulery, J. D., 1981, *Job Descriptions in Manufacturing Industries*, New York: Amacom (Division of the American Management Association).
US Civil Service Commission, 1956, *Tests for Blind Competitors for Trades and Industrial Jobs in the Federal Civil Service*, Washington, DC: US Civil Service Commission.
US Civil Service Commission, 1978, Adoption by four agencies of uniform guidelines on employment selection procedures. Equal Employment Commission, Civil Service Commission, Department of Labor, Department of Justice, *Federal Register*, **43**, 38290-315.
US Department of Labor, 1972, *Handbook for Analyzing Jobs*, Government Printing Office No. 2900-0131, Washington, DC: Manpower Administration, U.S. Department of Labor.
US Department of Labor, 1977, *Dictionary of Occupational Titles*, Vol. 2, *Occupational Classification*, Washington, DC: Manpower Administration, US Department of Labor.
Vandergoot, D., 1982, Work readiness assessment, *Rehabilitation Counseling Bulletin*, **26**, 84–7.
Wagner, R., 1985, Job analysis at ARBED, *Ergonomics*, **28**, 255–73.
Weed, L. L., 1969, *Medical Records, Medical Education and Patient Care*, Chicago: Year Book Publishers.
Wehman, P., 1986, Supported competitive employment for persons with severe disabilities, *Journal of Applied Rehabilitation Counseling*, **17**, 24–9.
Welford, A. T., 1965, Performance, biological mechanism and age: a theoretical sketch, in *Behaviour, Ageing and the Nervous System*, edited by A. T. Welford *et al.*, Springfield, IL: Charles C. Thomas.
Whelchel, B. D., 1985, Recruitment: use performance tests to select craft apprentices, *Personnel Journal*, **64**, 65–99.
WHO, 1980, *International Classification of Impairments, Disabilities and Handicaps*, Geneva: World Health Organization.
Wygant, R. M., 1983, Job analysis: a review of concepts and methods. *Proceedings of the Annual Conference of the Institute of Industrial Engineers, Norcross, GA.*

Glossary

Acceptability scaling A technique of effort rating in which a subject is invited to adjust a workload, for example the weight of a lift under controlled circumstances, to a level which he/she considers to be the maximum tolerable.

Accommodation In human resource management, the requirement on the part of an employer to so modify the work or workplace that it is suitable for a handicapped or disabled employee.

Acromion The flat projection on the outer edge of the scapula that forms the point of the shoulder.

Aerobic In the presence of oxygen; applied commonly to that part of carbohydrate metabolism which utilizes oxygen.

AET Acronym for Erhebungsverfahren zur Tatigkseitanalyse; refers to an ergonomically based method of physical demands analysis.

Anaerobic In the absence of oxygen; applied commonly to that part of carbohydrate metabolism in which oxygen is not used.

Anthropometry The study of human dimensions, e.g., height, weight, arm length, and so on.

Arrythmia Irregularity of the heart beat.

Audiogram A record of hearing capacity presented in the form of a graph of sound frequency in hertz and hearing threshold in decibels at that frequency.

Audiometry The measurement of hearing threshold.

Auscultation The act of listening to sounds within organs, e.g., the heart.

Benchmark job Job which can be used as a yardstick against which other jobs can be compared.

Benesh notation A technique of chroeographic recording using a matrix to represent the human body. Ledger lines are used to notate extensions above and below the body. Symbols represent lateral and fore and aft movements while vertical direction is indicated by position of the symbol on a five line stave.

Biomechanics Study of the mechanics of human (or animal) motion, with particular respect to the functional anatomy and physiology of muscles and joints.

Cardiac output The timed volume of blood pumped by the heart, e.g., $ml\ min^{-1}$.

Cardiotachometer A device for recording the frequency of the heart beat (e.g., beats per minute).

Carotid pulse The arterial pulse as palpated at the carotid artery in the neck.

Category scaling A concept based on a presumed relationship between perceived intensity and effort. No attempt is made, however, to compare with a known standard or estimate a ratio. In this situation each end of the scale is defined. For example, 1 might be the level at which effort (or pain) is first perceived; 5 might be when it become intolerable. Other levels are dispersed in between.

Cerebral palsy Disability resulting from damage to the brain before or during birth; may result in muscular incoordination and speech disturbance.

Cirrhosis Damage to, and destruction of, the liver, accompanied by excessive formation of connective tissue. Often associated with continued or repeated ingestion of toxic materials.

Cluster analysis Statistical technique used to associate commonalities in a population of items.

Decibel Unit of sound measurement representing one-tenth of a Bel, and abbreviated as dB. 0 dB, by definition, is the threshold of hearing; 35 dB the noise expected in a quiet library, and 100 db the sound of a very noisy factory.

Demographic Pertaining to the statistical study of populations, for example size, density, and vital statistics.

Diastolic blood pressure That pressure in the blood circulation which is found during the relaxation, or diastolic, phase of the heart contraction cycle.

Dioptre Unit of measurement of the strength of a lens; reciprocal of the focal length in meters.

Dynamic flexibility Contrasted, in the work of Fleishman and his colleagues on effort, with Extent flexibility; involves the capacity to make repeated flexing movements, for example, rapidly rotating joints or body parts.

Dynamic work In ergonomics and work physiology, work that occurs when a task requires repeated, commonly rhythmic, contraction and relaxation of muscle, as for example in moving a lever to and fro, rather than fixed holding.

Dynometer Hand-held device for measuring grip strength; the handles are squeezed together while the exerted force is recorded on a meter integral with the device.

Elemental motion In motion-time-study (MTM), a specific motion occurring in the course of a task.

Endocrine Pertaining to the group of ductless hormone-producing glands in the body, or to the hormones themselves.

Ergometer Device, commonly in the form of a stationary bicycle, or a series of steps of standard height, used in work physiology to provide a controlled work output.

Ergonomics The scientific study of the relationship between the person and his/her working environment; encompasses elements of human anatomy (anthropometry), human psychology, human physiology, and engineering.

Eskol-Wachman notation Means of recording human motion developed in choreography. Based on consideration that movements around joints describe spheres. Movements defined by way of end-point coordinates and the distance from the end-point to the centre of the relevant sphere.

Expiratory reserve volume In pulmonary physiology, the volume of air that can be expired by a forced maximum expiration after completion of a normal expiration.

Explosive strength In the work of Fleishman and his colleagues on effort, the ability to expend the maximum amount of energy in activities involving one or more maximum thrusts.

Extent flexibility In the work of Fleishman and his colleagues on effort, the ability to extend the trunk, arms, and/or legs through a range of motions.

Fat calipers Hand-held device for measuring the thickness of a skinfold. Used to estimate the depth of human body fat.

Flexometer A two-armed protractor for measuring the extent of flexion, particularly at the hip.

Forced expiratory volume In pulmonary physiology, the maximum volume of air that can be exhaled over a period of one second.

Functional capacity The ability of a person to perform the essential duties of a job.

Functional job analysis The process, according to the work of Fine, of analysing jobs using standardized terminology to generate formal task statements to describe a job in relationship to its requirements for data, people, and things.

Functional residual capacity In pulmonary physiology, the volume of air left in the lung after completion of a normal expiration.

Gas analysis In pulmonary physiology, the process of analysing breathing air for its content, in particular, of oxygen and carbon dioxide.

Gluteal line The skin surface marking outlining the edge of the gluteal muscle.

Glycogen The main source of stored carbohydrate in the body.

H-AET An extention of AET which takes into account anthropometric limitations and activity components that place demands on the worker.

Haemic Pertaining to the blood.

Human factors engineering The scientific study of person–machine relationships; concerned largely with determining the requirements of engineering design for human use.

Inspiratory capacity The volume of air that can be inspired above the level of the functional residual capacity (that is, after completion of a normal inspiration).

Isometrics Exercise in which opposing muscles are so contracted that there is little shortening of the muscle fibres involved but great increase in their tone.

Job matching Comparison of the functional capacity of a worker with the physical demands of a job, with the objective of determining the suitability of a job for a worker or of selecting the most appropriate worker for a job.

Kilopond The force acting on the mass of 1 kg at 1 g.

Kinesiology The scientific study of the principles of mechanics and anatomy in relation to human movement.

Labanotation Method of recording human motion in choreography on a staff containing columns corresponding to different body party. Direction in lateral and fore and aft directions is coded by the shape of the symbols while vertical directions are coded by symbol shading. Time is noted along the vertical axis.

Locomotor Pertaining to those human physiological systems concerned with motion.

Lumbosacral disc Fibrocartilagenous disc separating the lumbar from the sacral vertebrae; prone to injury during heavy or repeated lifting.

Lymphatic Pertaining to the lymph system, that is, that system within the human body responsible for drainage and circulation of tissue fluid.

Maximum voluntary ventilation Maximum volume of air that can be breathed in a 15 second period.

Metabolic rate A measure, in kilocalories, of the energy required for human activities.

Methods and task analysis Generic term for industrial engineering techniques largely concerned with measurement of time taken to perform tasks and/or task components.

Morbidity data Statistics pertaining to sickness and injury.

MPSMS Term used in the USES method of job analysis, referring to materials, products, subject matter, and services.

Muscular dystrophy (musculodystrophy) Neurological illness characterized by progressive weakness of muscles.

Musculoskeletal system Generic term for the system of bones, joints, muscles, and their appurtenances.

Nomogram Graph that permits, by the use of a straight edge, a reader to determine the value of a dependent variable when two or more independent variables are given.

Octave (octave band) In sound measurement, the series of tones or frequencies comprising an interval between two frequencies, one of which is double the other.

Oculomotor Pertaining to the muscles of the eye.

Olecranon One of two lateral bony projections at the lower end of the humerus bone (that is, at the elbow).

Oxygen uptake The volume of oxygen extracted from inspired air during human activity.

Paradigm A model or example.

Phoria Generic term for a variety of latent forms of imbalance in the optical axes of the two eyes.

Physical abilities analysis In job analysis, the attempt to quantify by scaling methods the perception of effort or exertion experienced while doing a task.

Plethysmograph Electronic device used for measuring, for example, change in finger volume and thereby allowing calculation of blood flow.

Pneumotachograph Electronic device used in pulmonary physiology for measuring flow of expired air.

Position strength Maximum voluntary static strength that can be exerted at a point in the work envelope at which a given weight is maintained.

Posture coding Term applied to a procedure for defining body posture and position by representing the body as having a group of targets around limbs and trunk. To represent departures of the body part from the standard position in the horizontal plane a mark is made on a target according to its radial displacement, the straight-ahead position being vertically upwards on the target. Displacement in the vertical plane is recorded by treating the concentric circles as a scale of 45, 90, and 135 degrees respectively from the target centre.

Pre-employment examination A medical examination, normally conducted by a physician, to determine the medical status of a job applicant before he is hired.

Pre-placement examination A medical examination, normally conducted by a physician, to determine the general suitability of an applicant before he is placed in a particular job. It is neither job, nor task, specific.

Pulmonary function testing Term applied to a range of test procedures used to determine the functional capacity of the respiratory system.

Pulmonary membrane Anatomical term describing the microscopic barrier through which air and blood are exchanged in the lung; comprises the alveolar (air sac) wall, the interstitial fluid (i.e., the microscopic layer of fluid between the alveolar wall and capillary wall), and the wall of the capillary (microscopic blood vessel) serving that alveolus (air sac).

Radial pulse Arterial pulse as felt at the radial artery at the wrist.

Rating of Perceived Exertion (RPE) Pseudoquantitative rating scale of 15 values, only some of which are defined, describing the magnitude of perceived exertion during various forms of exertion. Using this scale the severity of exertion has been shown to reflect increase in heart rate.

Ratio scaling Pseudoquantitative method of defining the functional relationship between perceived magnitude and physical dimensions. The subject under test estimates the percentage magnitude of a stimulus as compared with a known standard.

Residual volume In pulmonary physiology, the air remaining in the lung after a forced maximal expiration.

Sagittal The median plane of the body, or any plane parallel to it.

Scapulodorsal The region of the shoulderblade.

Skinfold A pinch of skin selected for use with fat calipers to determine the depth of body fat.

Sphygomanometer Device for the manual estimation of human peripheral blood pressure.

Spirometer Device for measuring lung volumes and capacities.

Splenomegaly Enlargement of the spleen.

Static strength In the work of Fleishman and his colleagues on effort, the force that a person exerts in lifting, pushing, pulling, or carrying external objects. It represents the maximum force a person can exert for a brief time period where the force requires some major effort. Because it involves a brief time period resistance to fatigue is not a feature as in dynamic strength.

Steady state conditions In human physiology, the state in which human body function, as represented for example by heart rate, blood pressure, breathing rate and so on, have reached equilibrium for a given exertion.

Systolic blood pressure Peak blood pressure found during the contraction, or systolic, phase of the cardiac contraction cycle.

Task element Componet of a task, particularly in motion-time analysis.

Task inventory A form of job evaluation questionnaire comprising a listing of tasks within some occupational field, along with an indication of their occurrence in the work of the incumbent and the relative amount of time spent on each.

Taxonomy Orderly classification according to presumed natural relationships.

Tidal volume In pulmonary physiology, the air moved in and out of the lungs during each breathing cycle.

Time and motion study Also known as methods-time measurement; a proprietary technique of describing a task in terms of the human motions required to undertake it, such as reach, move, turn, apply pressure, grasp, and so on, along with the time taken for each motion.

Total lung capacity The sum of the functional residual capacity and the inspiratory capacity.

Trunk strength In the work of Fleishman and his colleagues on effort, trunk strength is a derivative of dynamic strength and is characterized by the resistance of trunk muscles to fatigue over repeated use. It describes the degree to which the abdominal and lower back muscles can support part of the body repeatedly or continuously, as in obtaining or recovering from postures involving flexion at the hip, such as leg-lifts or sit-ups.

USES An acronym for a programme developed by the United Stated Employment Service of the Training and Employment Administration (USES) as presented in the *Handbook for Analysing Jobs*. The USES job analysis record is presented on a standardized form outlining job information in predetermined standardized terms which are specified in the previously noted Handbook for Analysing Jobs. The concept of the analysis is based on defining the extent of involvement of the worker in three areas, namely, data, people and things.

Ventilation minute volume The volume of alveolar ventilation (i.e., the air entering the air sacs), times the frequency of breathing, times the volume per breath (the tidal volume) minus the volume of the dead space (i.e., that space in the respiratory passages not involved in gas exchange).

Ventilatory capacity Ability to inhale and exhale, a property dependent on the mechanical characteristics of the chest wall and the lung (e.g. compliance and resistance).

Work capacity The integrative sum of the fucntions of the respiratory, cardiovascular, and musculoskeletal systems. It can be defined in terms of the:

(a) abililty to breathe
(b) ability to transfer oxygen effectively in the lungs
(c) ability to increase cardiac output to meet a given work load
(d) ability to transport oxygen to the working muscles
(e) ability to exert adequate muscular force.

Worker function scales Scales used in Functional Job Analysis whereby the job is analysed in terms of the interaction of the worker with the three hierarchies of analytical terms, namely, data, people and things.

Index

absenteeism 114,
acceptability scaling 70
accommodations 99
  application 102
  costs 101
  individualization 100
  legislation 100
  requirements 100
action for limit for lifting 64
activity analysis 48
adiposity 90
adverse impact theory 120
aerobic fitness testing 89
AET approach 44
aircraft assembly 113
American Medical Association 4, 13
anthropometry
  fitness testing 88
  limitations 48
Arbeitswissenschaftliches Ehrebungsverfahren zur Tatigkeitsanalyse (AET) 44
architectural barriers 104
arm motion recording 61
audiometry 93
average work output 66

barriers 43
  removal 104
behaviour description 67
behaviour requirements 67
Benesh system 61
bicycle ergometer 165
Biographical Data questionnaire 74
biomechanical techniques
  work capacity measurement 165
biomechanics 173
  of task demands 63
blind workers 105
body co-ordination 71
body segment motions 61, 65
body weight 90
Borg scale 69
business necessity
  selection criteria 121

Canadian Standardized Test of Fitness 86
cardiopulmonary function 165
carrying recommendations 63
category scaling 69
causes of rejection 19
checklists
  Essential PDA 58
  visual demands 43
  work activities 4
Civil Rights Act 120
classification
  job evaluation 25
  job matching 108
  medical fitness 108
  of disabilities 4
  of health status 16
  physcial job demands 48
Classification Act 25
clinical methodology
  physical examinations 13
cluster analysis of task components 49
coding
  job matching 108
  of motion 61
  of posture 60
  procedures 53
confidentiality 20
content validity of examinations 121
costs for accommodation 101
criticality of tasks 41

data acquisition and recording
  job analysis 40, 41, 59
deaf workers 106
definitions of terms
  in disability 6
  in job analysis 24
direct analysis of task components 48
direct observation of the worker 40
disability
  definition of terms 6
  rating methods 5
  scales 16
Disabled Persons (Employment) Act 119
disabled workers 42, 99
  legal considerations 119
  testing 104
discrimination
  by employers 3
  by sex 121
  of the handicapped 119
dishwashing 125
disparate treatment 120
dynamic flexibility 71
dynamic strength 71

effort perception
  physical factors 70
  quantification 68, 72
employment
  discrimination 120
  equity 99
  of the handicapped 1, 119
endurance 89
environmental conditions 43
Eskol-Wachmann method 61
Essential PDA 57
  checklist 58
explosive strength 71
extent flexibility 71

fat distribution 90
FCA *see* functional capacity analysis
fitness-for-work examinations 12
fitness testing 86, 90
flexibility 71, 89
functional ability evaluation 8
functional capacity assessment 8, 81
functional job analysis 29
functional ratings 112

gait assessment 61
gas analysis of expired air 167
grade description system 25
graphical systems
  data recording 59
guidelines
  for human rights 100
  for physical examinations 13, 19, 20
  for selection procedures 121
  for task statements 37
GULHEMP method of job matching 112, 183

H-AET supplement 48
handicapped workers 42, 99
  legal considerations 119
  testing 104
Hanman method of job matching 111
hardship
  accommodation for the disabled 100
health and safety
  accommodation aspects 102
health histories 14, 82
health status classification 16
hearing impaired 106
hearing testing 92
heart rate 169
history of screening methods 1
Human Rights Codes 8
human rights guidelines 100
human stress 44, 47

illnesses/injuries 82, 114
institutional barriers 104
insurance schemes 3
interviews 40, 47

Job analysis 23
  application 72

data acquisition 40
definitions 24
legal aspects 121
NEMA system 35
physical demands 42
USES 26
job demands 39
analysis 45
biomechanics 63
job dendogram 49
job evaluation 23
qualitataive approaches 24
quantitative appraoches 34
job factors 34
job matching 1, 7, 81, 85, 107
history 107
practice 107
qualitative approaches 109
quantitative approaches 111
theory 107
validity 113
job modification 103
job profile method 53
job profile rating scales 55
job related stress
selection criteria 120
job requirements 108
job strength rating 175
job structuring 103
job task listing 43

kitchen workers 125

Labanotation 61
legal aspects 66, 119, 121
legislation 1, 2, 100, 119
lifestyle history 84
lift strength ratio 174
lifting 174
demands model 64
limits formula 64
postures 64
recommendations 63
lung function 94

magnitude estimation 68
manual handling
effort 56
postures 57
maximal oxygen uptake 165
maximum aerobic power 91, 165
maximum permissible limits for lifting 64
medical categories 16
medical examinations 11, 14, 82, 84
development 4
origins 2
medical fitness 81
medical history
general 83
medical records 20
mental health 17
models
lifting demands 64
task performance 115
motion coding 61
motion of body segments 61, 65
motor impaired workers 106
muscular strength 43, 71, 89, 173

NEMA job analysis system 35
nomogram for cardiopulmonary function 168
notational systems for data recording 59, 61

observation of workers 40, 47
occupational health history 82
occupational qualification
selection criteria 121
Ontario Human Rights Code 100

PAAM *see* Physical Abilities Analysis Manual
PDA *see* Physical Demands Analysis
perceived exertion 68, 72
perception of effort
physical factors 70
quantification 68, 72
periodic examinations 12

photography 59
physical abilities analysis 72
Physical Abilities Analysis Manual (PAAM) 73, 77
physical barriers 43
Physical Demands Analysis 8, 39, 42, 67, 81, 177
physical demands job analysis 42, 125
physical demands principles 39
physical examinations 11, 84
physical fitness testing 86, 90
physical health 17
physical workload 169
physically handicapped workers 42, 99
  legal considerations 119
  testing 104
physiological functions 85
physiological techniques
  work capacity measurement 165
pictograms of common postures 54
points rating 34
postural stress 55
posture coding 60
posture recording 54, 60
posture targetting 60
posturegram 60
predetermined ratings 25
pre-employment examinations 12
pre-placement examinations 12, 82
psychophysiological methods 175
PULHEMS 108, 112

questionnaires
  Biographical Data 74
  health histories 14
  job evaluation 36
  physical abilities 73
  task demands 43
  WCAM Rating 74
  work system evaluation 47

ranking in job evaluation 25
rating of perceived exertion (RPE) scale 69
rating scales 5, 53, 69, 112
  for critical task dimensions 41
  PAAM 77
  PAAM/WCAM 73
  WCAM/WPAM 159
rating/scaling 47, 78
ratio scaling 68
Rehabilitation Act 100
rejection causes 19
reliability of PDA data 42
restricted placement 84
restrictions 99
risk assessment 84

selection procedure guidelines 121
sex discrimination 121
sickness 82, 114
simulation of accident/illness behaviour 117
site modification 103
social conditions 43
social insurance 3
social reform 2
stamina 71
static strength 71
static work stress 47
step testing 167
strain 44
strength 43, 71, 89, 173
stress 44, 47, 70
  postural 55
  work effort 55
subjective techniques 68
suitability of candidates 17
support services 104
symbolic notation 61
system analysis 45
systems review 85

task analysis 41, 43, 45, 72
task banks 41
task criticality 41
task demands
  biomechanics 63
task inventories 36
task performance model 115

task statements 30, 37
taxonomies 48
testing
  fitness 86, 90
  of the handicapped 104
treadmill 165
trunk strength 71

USES job analysis 26

validity of PDA data 42
video recording 59
vision testing 92
visual demands checklist 43
visually impaired workers 105

War Manpower Commission 4
WCAM *see* Working Conditions Analysis Manual
WCAM/WPAM analyses 159
WCAN Rating questionnaire 74
work capacity 81, 85, 177
  mesurement 165
work effort stress 55
work output modelling 66
work restrictions 18, 102
work systems evaluation 44
workers' compensation 3
Working Conditions Analysis Manual (WCAM) 73
Workmen's Compensation Act 3
World War II effects 4

X-ray examinations 15
  of the back 16

For Product Safety Concerns and Information please contact our EU representative GPSR@taylorandfrancis.com
Taylor & Francis Verlag GmbH, Kaufingerstraße 24, 80331 München, Germany

www.ingramcontent.com/pod-product-compliance
Ingram Content Group UK Ltd.
Pitfield, Milton Keynes, MK11 3LW, UK
UKHW021612240425
457818UK00018B/526

* 9 7 8 0 8 5 0 6 6 8 5 8 2 *